THE
PROGRESSING CAVITY PUMP
HANDBOOK

THE
PROGRESSING CAVITY PUMP HANDBOOK

James M. Revard

PennWell Books

Copyright 1995 by
PennWell Publishing Company
1421 South Sheridan/P.O. Box 1260
Tulsa, Oklahoma 74101

Library of Congress Cataloging-in-Publication Data

Revard, James M.
   The progressive cavity pump handbook
     p. cm.
   Includes bibliographical references and index.
   ISBN 0-87814-445-5
   1. Pumping machinery — Handbooks, manuals, etc.
TJ900.R49   1995
621.6'9—dc20

All rights reserved. No part of this book may be reproduced, stored in a retrieval system, or transcribed in any form or by any means, electronic or mechanical, including photocopying and recording, without the prior written permission of the publisher.

Printed in the United States of America

1 2 3 4 5 99 98 97 96 95

To my wife, Dianna,

who stood by me and believed in me

while completing this book.

You make the work worthwhile.

A special dedication to

the greatest oilman that I have had

the privilege of working with—

Papoose, your tick lives on.

Thanks, Dad!

CONTENTS

FOREWORD viii

ACKNOWLEDGEMENTS x

HOW YOU WILL BENEFIT FROM THIS BOOK xi

CHAPTER 1   THE PROGRESSING CAVITY PUMP;
BEGINNING, HISTORY, AND DESIGN 1
PC History 1
Design 3
Materials of Construction 7

CHAPTER 2   THE BENEFITS AND ADVANTAGES OF
UTILIZING THE PROGRESSING CAVITY AS AN
ARTIFICIAL LIFT METHOD 11
Efficiency 12
Operating Cost 14
Capital Cost 20
System Reliability 24
Application Flexibility 26
Environmental Benefits 27

CHAPTER 3   THE LIMITATION AND DISADVANTAGES OF
THE PROGRESSING CAVITY PUMP AS AN
ARTIFICIAL LIFT METHOD 29
Depth and Volume Limitations 29
Temperature Limitations 31
Application Flexibility Limitations 33
Service and Experience Limitations 37
Salvage Value Limitations 38

CHAPTER 4   THE PROGRESSING CAVITY PUMP
DOWN HOLE APPLICATIONS 41
Common PC Applications 41
Unusual PC Applications 44
Screening and Sizing PC Applications ...51
Selecting a PC Pump Supplier 54

CHAPTER 5   INSTALLING THE PROGRESSING CAVITY PUMP
DOWN HOLE 61

	Determining Pump Setting Depth 61
	Installation of the Stator and Rotor 63
	Drive Assembly Installation 67
	Pre-Start Check List 71
CHAPTER 6	OPERATION AND MAINTENANCE OF THE PUMPING SYSTEM .73
	PC Pump Operating Performance 73
	Drive Assembly Operation and Maintenance 76
CHAPTER 7	TROUBLE SHOOTING THE PROGRESSING CAVITY PUMP 79
	Trouble Shooting Problems 80
	Common Failures .87
CHAPTER 8	PRIME MOVERS AND OTHER ACCESSORIES FOR THE PROGRESSING CAVITY PUMP 97
	Prime Movers 98
	Sucker Rod Guides 104
	Anchor-Catchers .108
	Gas Separators 109
	Flow Controls/Pump Off Controls 111
	Anti-Friction Unit 117
CHAPTER 9	THE IMPORTANCE OF PUMP PERFORMANCE TESTING **121**
	Initial Performance Testing 122
	Pump Inspection and Retesting 127
CHAPTER 10	MEETING THE FUTURE ARTIFICIAL LIFT NEEDS WITH THE PROGRESSING CAVITY PUMP . . .**135**
	Industry Needs 136
	Future Developments 140
	ISO 9001 146
CHAPTER 11	CONCLUSIONS **149**
REFERENCES	152
INDEX	153

FOREWORD

The idea for this book was partly the result of a very well written SPE paper published by J. D. Clegg, S. M. Bucaram, and N. W. Hein (SPE 24834) in which they performed a comparison of artificial lift methods including the Progressing Cavity Pump.

In that SPE paper, the authors suggest that other professionals publish new data to improve the industry knowledge on artificial lift selection. My purpose in writing this book is to help educate the users on the application of the Progressing Cavity Pump. Many areas of the book will concur with the SPE paper by Clegg, Bucaram, and Hein, but will also discuss new technology and practices that have made it possible to respond to some of their concerns over PC pumps.

It is critical that the proper selection of artificial lift method be chosen to insure long-term profitability of most producing wells. The wrong selection can lead to reduced production and increased operating costs.

This book will focus on one method of artificial lift, the progressing cavity pump (PC pump). The PC pump has become a widely accepted and utilized method of artificial lift. It offers numerous advantages over other artificial lift methods and has proven itself to reduce operating costs while maximizing production.

Through research and development of the progressing cavity pump design, the production and lift capabilities have expanded to cover a wide range of well conditions. The pump continues to prove itself to be more efficient and economical than other methods of artificial lift.

This book will cover all the benefits of the PC pump as well as its limitations. It will discuss the range of applications for the PC pump, the preparations needed to apply it, and its installation and operation. In addition, it will discuss the need for testing before, during, and after installing a PC pump.

In addition, the book will cover the future needs for the progressing cavity pump in the oil and gas industry — a "Where-do-we-go-from-here?" approach.

Currently, the research shows that there is a growing need in the oil and gas industry for a more efficient, cost effective way to produce oil and gas—the progressing cavity pump provides the way.

ACKNOWLEDGEMENTS

Bach Oilfield Services, Inc. for their assistance in providing expertise in regards to performance testing and applying the PC pump, as well as their assistance in artwork and photos.

Griffin/Legrand for their assistance in the use of their printed manuals and photos

HOW YOU WILL BENEFIT FROM THIS BOOK

The main purpose of this book is to assist the oil and gas producer in achieving a successful installation and operation of the Progressing Cavity Pump. At the same time, a related purpose is to insure the continued expansion of the PC pump marketplace by educating the reader so as not to make some of the mistakes that have been made by manufacturers, pump suppliers, and producers in the past.

This book is designed to answer the questions that confront the oil and gas producer who has installed a Progressing Cavity Pump or is considering the utilization of the PC pump.

The reader will receive a brief history lesson on the PC pump as well as on the design and materials of construction.

This book will also provide the reader with detailed information on the benefits of utilizing a Progressing Cavity Pump in a well, such as capital cost savings, higher efficiencies, and lower operating cost.

The oil and gas producer will also benefit by learning the limitations of the PC pump in order to avoid the misapplication of the pump product. Common PC pump applications will be discussed as well as unusual applications.

The reader will learn the proper procedure for screening a pump application and how to go about selecting a pump supplier.

The book will cover the nuts and bolts of the installation of a Progressing Cavity Pump, from determining the pump-setting depth to a pre-start check list.

The field foreman or supervisor will find down-to-earth advice on common failures of the Progressing Cavity Pump, how to identify problems, and suggestions and procedures for solutions.

This book is a quick reference guide for petroleum engineers, independent producers, field foremen or supervisors, and pumpers alike to find helpful information on the overall operations of the Progressing Cavity Pump that will insure its successful installation.

By utilizing this book, all of this information and much more is at the oil and gas producer's fingertips. This is an easy-to-use, down-to-earth reference manual that will answer many questions the oil and gas operators have about the Progressing Cavity Pump.

CHAPTER 1

THE PROGRESSING CAVITY PUMP: BEGINNING, HISTORY, AND DESIGN

PC History

The PC pump has a long history. It was invented and designed in the late 1920s by the Frenchman René Moineau.

Moineau set out to design a rotary compressor and in the process created a new rotary mechanism to be used for the utilization of variations in the pressure of a fluid, which he referred to as *capsulism*. His goal was to make it possible to use this capsulism in pumps, compressors, or motors.

In the early 1930s, the Progressing Cavity Pump's principle patent was licensed to three companies: PCM-Pompes of France, Mono Pumps LTD of

England, and Robbins & Myers, Inc. of the United States. Over the years, other small pump companies have manufactured spin-offs of the Moineau principle.

The Moineau principle has been utilized in many industries in a wide variety of applications since its licensing. It has been used as a pump in just about every industry: chemical, coal, food, metal working, mining, paper, petroleum, textile, tobacco, and water and waste water treatment. In the petroleum industry, the PC pump has been utilized as a surface transfer pump for over 50 years.

In the mid-1950s, the progressing cavity pump principle was applied to hydraulic motor applications by reversing the function of the progressing cavity pump. The device was then being moved by fluid instead of pumping fluid. With the pump elements being driven by drilling mud or other fluids, it became the prime mover for drill motors. The Moineau principle is now being widely used in the petroleum drilling industry.

Then in the early 1980s, the PC pump was utilized as an artificial lift method in the petroleum industry. Robbins & Myers, Inc. of the United States have to be credited with being the first to apply the Moineau principle to artificial lift in the petroleum industry. They became the first PC pump manufacturer to market the pump as an alternative to conventional lift methods and to establish a new marketplace for the PC pump. Since the mid-1980s, other manufacturers have entered into this marketplace, expanding the acceptance of the product by the oil and gas industry.

The pump is applied to artificial lift by attaching the pump elements to the tubing and rod string. The stator is run on the end of the tubing, and the rotor is attached to the bottom of the rod string and landed in the stator. The rods and rotor are rotated through a wellhead drive assembly that is designed to carry the weight of the rod and the fluid column. This setup will be discussed in more detail in a later section.

Currently, the PC pump is being widely used for lifting fluids from depths of 6,000 ft. and deeper in oil and gas wells. The progressing cavity pump offers to the petroleum industry a great number of advantages over traditional lift equipment, of which the most important is lowering the cost per barrel lifted.

DESIGN

The Progressing Cavity Pump design consists of a single external threaded helical gear (rotor) that rotates eccentrically inside a double internal threaded helical gear (stator) of the same minor diameter and twice the pitch length (Figure 1-1). The rotor and stator form a series of sealed cavities, 180° apart, that progress from the suction to the discharge end of the

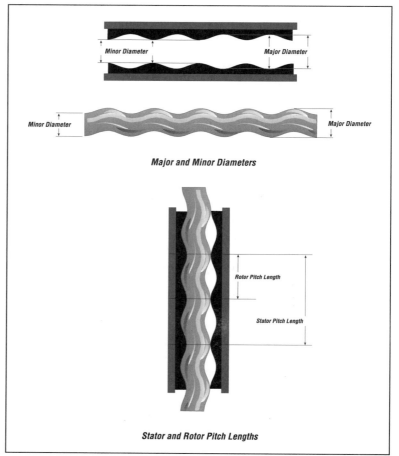

Fig. 1-1

pump as the single external threaded helical rotor rotates. As one cavity diminishes, the opposing cavity increases at the same rate, which keeps the fluid moving at a fixed flow rate that is directly proportional to the rotational speed, resulting in a pulsationless positive displacement flow. The total cross-sectional area of the cavities remain the same regardless of the position of the rotor in the stator, as shown in Figure 1-2.

The displacement of the PC pump in addition to

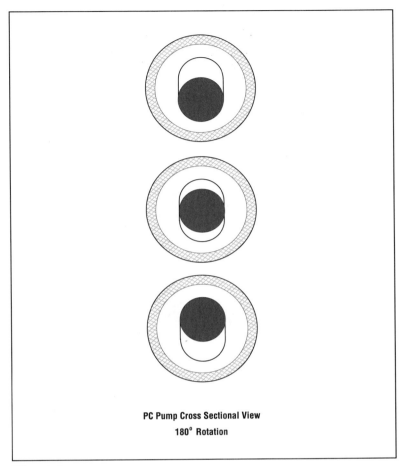

PC Pump Cross Sectional View
180° Rotation

Fig. 1-2

being a function of the speed, is directly proportional to three design constants: the cross-sectional diameter of the rotor, its eccentricity (or radius of the helix), and the pitch of the stator helix. The displacement per revolution will vary with the size of the cavity area (Figure 1-3).

The pressure capability in a progressing cavity pump is a function of the number of times the seal lines that are formed by the rotor and stator are

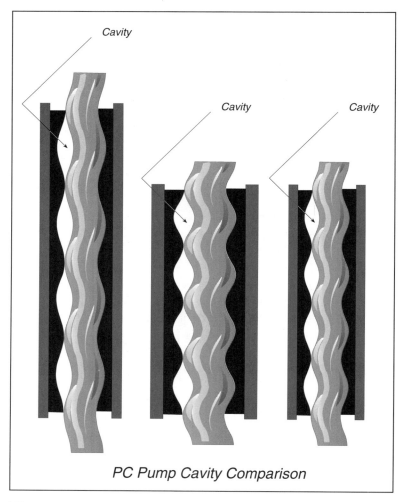

Fig. 1-3

repeated or the number of stages within the pump. Normally, a stage is designed to be about 1.1 to 1.5 times the pitch length of the stator. This insures a proper seal between the rotor and stator (Figure 1-4). The more seal lines or stages a pump has, the more the pressure capabilities increase, allowing a pump to perform at a greater depth in a well.

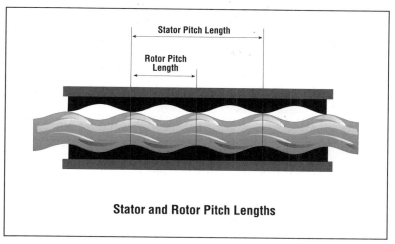

Fig. 1-4

As the pressure increases for a given number of stages and speed, the flow rate will decrease. This reduction in flow rate caused by the higher pressure is called slip.

All progressing cavity pumps experience slippage between the suction and discharge ends. The amount of slip is determined by the pressure and is independent of speed.

Three factors that will affect the amount of slip within a PC pump due to pressure are: the number of seal lines or stages, the viscosity of the fluid being pumped, and the compression fit between the rotor and stator.

Materials of Construction

The development of materials of construction for the progressing cavity pump is an ongoing effort by the manufacturers.

The stator consists of a steel tube with an elastomer that is permanently bonded to the inside and molded in the configuration of a double helix.

The external single helical rotor is normally alloy steel, machined to an exact tolerance, and then chrome plated. The rotor can be machined in stainless steel when needed for corrosive applications.

Great care and quality control during manufacturing of the rotor and stator insure that the user will receive the best performance possible from the PC pump. Each PC pump manufacturer performs this task in a different manner, which should be considered when selecting the manufacturer. Pump test performance is a very good indicator of quality, which will be covered in a later section.

As the progressing cavity pump becomes more widely accepted, the need for improvements in elastomer developments of the stator will increase in order to apply the pump to a greater range of producing wells.

Currently, there are three elastomer compounds on the market for use with the downhole progressing cavity pump, all of which are compounds formulated from nitrile rubber, also referred to as Buna N. Its copolymers are acrylonitrile and butadiene. The acrylonitrile content ranges from 20% to 50%. As the acrylonitrile content is increased, the oil and solvent resistance improves, its tensile strength increases, hardness increases, the abrasion resistance increases, the high temperature resistance improves, and its resilience and permeability qualities decrease.

Acrylonitrile content is usually the first consideration when designing a nitrile compound.

The three variations of the compound are:
- Medium-High Acrylonitrile
- Ultra-High Acrylonitrile
- Hydrogenated, Highly Saturated Nitrile (HSN)— Very High Acrylonitrile

Medium-High Acrylonitrile (ACN) Nitriles

The medium-high acrylonitrile elastomer has good oil and solvent resistance. This elastomer can be applied in oil of less than 28 API, and in application of high water cuts or 100% water. It has excellent abrasion and mechanical properties. The temperature limitation of this type of elastomer is 200°F. Of all the elastomers on the market, this one is best suited for CO_2.

Ultra-High (ACN) Nitriles

The ultra-high acrylonitrile elastomer has very good oil and solvent resistance. The higher ACN content gives this elastomer increased aromatic swell resistance. This elastomer can be applied in oil of 28 API to 38 API gravities. In oils of this type, a fluid sample should be tested to check the degree of aromatic swell. This elastomer also has good resistance to swell from treating chemicals; however, all chemicals should be tested with the elastomer prior to applying the products.

The mechanical properties and abrasion resistance is good. The temperature range of this elastomer is 225°F.

Highly Saturated Nitrile (HSN) Very-High Acrylonitrile

This elastomer resistance to aromatic swell is lower than that of the ultra-high (ACN) nitriles. The mechanical properties and abrasion resistance are good. The temperature range is the highest of the three elastomers at 275°F. This elastomer has fair resistance to hydrogen sulfide and good resistance to iron sulfide. Again, the well conditions should be examined closely before applying the product.

Formulating compounds is a fine art form, and it is very critical that the research is aggressively pursued.

CHAPTER 2

THE BENEFITS AND ADVANTAGES OF UTILIZING THE PROGRESSING CAVITY PUMP AS AN ARTIFICIAL LIFT METHOD

There are many benefits and advantages to utilizing a PC pump as a method of artificial lift, some of which include the following:
- Higher volumetric efficiencies.
- Utilization of smaller motors resulting in lower lifting cost.
- Capital cost is normally less than other artificial lift methods.

- Fewer moving parts, resulting in lower maintenance.
- Application flexibility.
- Environmentally pleasing/Low profile unit.

EFFICIENCY

The progressing cavity pump offers the end user an artificial lift method that is highly efficient and has been referred to as "excellent" by many artificial lift experts. (See SPE 24834 paper by J. D. Clegg, S. M. Bucaram, and N. W. Hein.)

Most PC pumps will have a higher efficiency rating than that of a beam pumping system or an electrical submersible pumping system.

The typical PC pump system will have an efficiency of 70–98%, while the beam pumping system's efficiency will be 50–60%, and the electrical submersible pumping system efficiencies will be 40–50%.

One of the great benefits to the PC pump is the fact that with most manufacturers, if the supplier is experienced and has the proper knowledge, he achieve whatever efficiencies is desired by changing the fit between the rotor and stator.

An end user should look to a supplier who has the test facilities to match the rotor and stator to his well conditions. By testing a stator with different diameter rotors, the supplier is able to achieve the maximum volumetric efficiencies with the lowest internal friction or torque possible. This results in a pump that will give the greatest volumetric efficiencies with very low HP requirements.

Any manufacturer of a PC pump is capable of achieving high volumetric efficiencies; however, only a few can do so without increasing their internal fit of the pump to a point that the torque becomes a problem, especially at low speeds. This brings us to another important point about the efficiency of PC pump, which will be covered in more detail in Chapter 6.

The efficiency of a PC pump is greater at higher speeds than at lower speeds. However, an operator has to consider the fact that at higher speeds there is increased wear on not only the pump itself but also on the sucker rods and tubing string. It is better to give up a little efficiency by running the pump at slower speeds to achieve longer run times and less wear on the pump, sucker rod, and tubing string.

Where possible, it is suggested that the end user install a pump that is slightly oversized for the application so he is able to operate the pump at a lower speed while achieving the desired production rate. This is another reason for selecting a supplier who runs performance tests on the pumps, and matches the rotor and stator to the operator's well conditions. Ideally, the operator wants a PC pump that will yield a high efficiency at a moderate speed (250 to 400 RPMs).

The operator will benefit not only from lower electrical consumption or HP requirements but also will achieve long pump life and experience fewer rod and tubing problems, resulting in lower operating costs.

OPERATING COST

In the SPE paper by Clegg, Bucaram and Hein, they describe the Progressing Cavity Pump's operating cost as potentially low, but short run life on the stator and rotor were frequently reported.

Prior to the fall of 1992, this was the case; however, since that time some pump suppliers have used the matched or performance fitted pump technology, which results in greater run times on the rotor and stator.

In all cases where I compared the operating cost of the PC pump to the beam unit and Electric Submersible pumps, the PC pump always had the lowest operating cost.

The largest portion of a well's economical picture is its lifting cost. The biggest advantage of the Progressing Cavity Pump over other artificial lift methods is by far its operating cost. This is due to its high efficiency design. With a PC pump, you rotate the rod string as opposed to lifting it with a conventional beam unit, resulting in the use of less horsepower.

Like the capital cost comparison of the Progressing Cavity Pumps to pumping units and electric submersible pumps, the greatest savings are in applications that call for high volumes and deeper depths. With the competitive environment the oil and gas industry is facing today, the operating cost is a large part of the economics of producing a field.

For confirmation of power savings you may wish to refer to an SPE paper written by another industry professional on the subject of efficiencies (see SPE 25448 by K. J. Saveth).

To confirm the power consumption savings that the PC manufacturers claim, I performed a series of field studies in three different fields. It is important to note the "field study" was a look at actual field conditions as they existed in each operator's field. It is not implied that the artificial lift methods were being optimized. In fact, in the case of the beam pumping units I found that some were not balanced properly.

During the study, I collected readings on each well's amperage,

CASE #1 - TABLE 2-1
PRODUCTION DATA

	Beam	ESP	PC
Total Depth (ft.)	4,799	4,933	5,029
Dynamic Fluid Level	4,521	4,588	4,800
Current Production	359	395	299
BWPD	347	384	289
BOPD	12	11	10

POWER READINGS

	Beam	ESP	PC
Average Meas. Amperage	81.9	16.0	21.9
Average Meas. Voltage	481.5	1,400	485
KW/HR	48.1	32.3	11.3
KW/Bbl	0.7	0.42	0.19
Cost/BBL oil @.055 KW	$5.40	$3.85	$1.50

voltage, kilowatt hours, current production rate, and dynamic fluid level.

In Case #1, I compared three wells: a beam pumping unit, an electric submersible pump and a Progressing Cavity Pump. In Case #2, the comparison was performed in a different field with two beam pumping units and two Progressing Cavity Pumps. In Case #3, I performed a comparison on a total of five wells, two beam pumping units, two electric submersibles and one Progressing Cavity Pump.

CASE #2 - TABLE 2-2
PRODUCTION DATA

	PC	Beam	PC	Beam
Total depth (ft.)	2,482	2,347	2,257	2,259
Dynamic Fluid Level	2,185	2,255	2,006	1,855
Current Production	815	465	1,930	1,247
BWPD	733	460	1,868	1,203
BOPD	82	5	62	44

POWER READINGS

	PC	Beam	PC	Beam
Average Meas. Amp.	39.5	32	52.2	71.6
Average Meas. Volt.	463.5	480	439.5	476.5
KW/HR	24.6	22.9	31.9	43.2
KW/Bbl	0.73	1.18	0.40	.83
Cost/BBL Oil @ $.055KW	$.40	$6.04	$.68	$1.30

The wells' perimeters are presented in Tables 2-1, 2-2 and 2-3, respectively, as well as the actual power readings and the cost/bbl oil @ $.055/kw.

To achieve the most fair and efficient means of having an apple to apple comparison, the figures were converted to kilowatts per one barrel at 1,000 feet of lift. The graphs in Figures 2-1, 2-2, and 2-3 help the reader better visualize the variations between each application.

The kw/bbl @ 1,000 ft. is the relationship of daily

CASE #3 - TABLE 2-3
PRODUCTION DATA

	Beam 1	Beam 2	ESP 1	ESP 2	PC
Total Depth (ft.)	4,220	4,200	4,300	4,400	4,200
Dynamic Fluid Level	3,813	2,238	3,215	3,681	2,737
Current Production	176	558	1,420	1,263	1,122
BWPD	165	537	1,398	1,244	1,104
BOPD	11	21	22	19	18

POWER READINGS

	Beam 1	Beam 2	ESP 1	ESP 2	PC
Avg. Meas. Amperage	16.8	35.3	54.0	64.6	19.1
Avg. Meas. Voltage	794	799	783	751	780
KW/HR	20.7	51.1	63.1	72.2	21.3
KW/Bbl	0.75	1.3	0.32	.38	0.18
Cost/Bbl Oil @ .055/KW	$2.50	$4.25	$3.75	$5.00	$1.50

power consumption to total daily production at respective dynamic fluid levels converted to 1,000 ft. of lift.

In all three studies, the benefits of the Progressing Cavity Pump was overwhelming based on power consumption savings. If an operator were to multiply the power savings by utilizing the Progressing Cavity Pump in every possible application in his field, he would experience a very large operating cost savings over a one-year period.

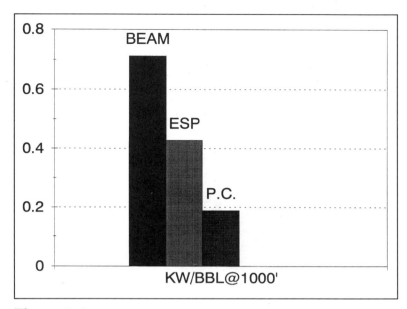

Figure 2-1

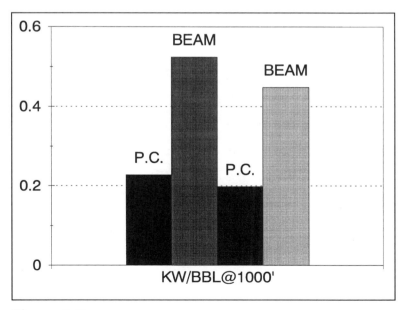

Figure 2-2

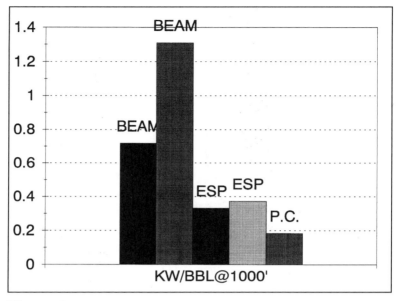

Figure 2-3

CAPITAL COST

For any producer, the capital cost of purchasing and installing lift equipment is a big concern. You have to take into account not only the cost of the lift equipment itself, but also site preparation, and the equipment necessary to install and set up the equipment.

In the majority of artificial lift applications that could utilize a progressing cavity pump, the capital outlay for the PC equipment will be less than that of a beam pumping unit system or an electric submersible system.

A few examples of this savings are that with a PC pump, you are utilizing smaller surface equipment, such as the drivehead electric motors and engines, there is no skid needed as with a beam pumping unit, no boom truck and additional workers needed to install or set up the surface equipment like a beam pumping unit.

As far as the savings between a PC and ESP is concerned, the capital cost savings is in equipment cost and the service rig time involved in the installation of the equipment.

The one area of applications in which I found the PC pump to be more expensive was in shallow depths with low volumes, i.e., 2,000 ft. or less with volumes under 40 BFPD. However, there may be a savings on electrical consumption.

When you start to approach deeper depths and larger volumes, the progressing cavity pump becomes the most economical means of artificial lift.

The following examples will show the capital cost savings an operator can expect by utilizing the PC pump. To try to have an apples to apples comparison, I included the cost of the tubing string on all three systems and the rod string on the PCP and beam unit.

EXAMPLE #1

Pump depth	2,000 ft.
BFPD	300
Fluid over pump	200 ft.
% water	90
Oil gravity	28 API
Casing size	5.5 in.
Tubing size	2.375 in.
Sand	None
H_2S	None
CO_2	None

CONVENTIONAL		ESP		PCP	
Equipment	List Price	Equipment	List Price	Equipment	List Price
114-143-64	$15,107.00	129 stage	$5,870.00	BDTX Drive head	$6,183.50
2-in. plunger	$1,558.00	19-HP motor	$8,500.00	20-1-095 rotor	$1,444.00
20-HP electric motor	$1,330.00	Round power cable	$4,300.00	20-1-095 stator	$1,901.00
Control panel	$2,326.00	Flat cable extension	$747.00	Frame and guard	$697.20
Polish rod	$99.35	Control panel	$4,202.00	10-HP elect. motor	$784.00
Polish rod liner	$234.75	$2^7/_8''$ tubing	$5,200.00	Control panel	$2,326.00
Tubing	$5,200.00			Polish rod	$99.35
3/4" rod string	$2,960.00			Tubing	$5,200.00
				7/8" rod string	$3,860.00
Total	$28,815.10		$28,819.00		$22,495.05

* Pricing based on manufacturer's suggested list price.

EXAMPLE #2

Pump depth	5,800 ft.
BFPD	1,000
Fluid over pump	200 ft.
% water	95
Oil gravity	32 API
Casing size	5.5 in.
Tubing size	2.375 in.
Sand	None
H_2S	None
CO_2	None

CONVENTIONAL		ESP		PCP	
Equipment	List Price	Equipment	List Price	Equipment	List Price
C-1280D-365-192	$81,337.00	203 stage	$10,858.00	B-DVX Drive head	$9,877.00
2-in. plunger	$1,558.00	76-HP motor	$19,200.00	60-1-195 rotor	$6,092.00
125-HP electric motor	$7,860.00	Round power cable	$14,384.00	60-A-195 stator	$8,034.00
Control panel	$2,326.00	Flat cable extension	$973.00	Frame and guard	$1,045.80
Polish rod	$132.79	Control panel	$6,623.00	40-HP elect. motor	$2,476.32
Polish rod liner	$234.75	$2^7/_8$" tubing	$5,200.00	Control panel	$2,326.00
Tubing 7/8" rod string	$5,200.00 $7,830.00			Polish rod Tubing	$99.35 $5,200.00
				1" rod string	$11,484.00
Total	$106,478.48		$57,238.00		$46,634.47

* Pricing based on manufacturer's suggested list price.

In example #1, the Progressing Cavity Pump is 21% less expensive than that of the conventional beam pumping unit and 22% less expensive than that of the electric submersible pumping system.

In example #2, the Progressing Cavity Pump was 56% less costly than that of the conventional beam pumping unit and 19% less costly than that of the electric submersible pumping system.

I was able to confirm the PC pump price advantage with a few major users of the pumps. In one field where the PC pump is being utilized at a depth of 2,500–3,000 ft. with volumes upwards of 1,000 BFPD, the operator is experiencing the following cost benefits:

COST RANGE

Progressing Cavity Pump	Beam Pumping Unit
$10,000 to $15,000	$25,000 to $40,000

These prices were less the cost of the rod and tubing strings. It is quite apparent that if an operator has an application where a PC pump can be utilized, he can benefit from a large capital cost savings.

SYSTEM RELIABILITY

The reliability of a progressing cavity pump system is much like that of traditional lift methods, i.e., proper design and practices will insure an excellent run time.

The experience of end users in regard to reliability has varied for many reasons. The most common problem in my opinion is the lack of experience of both the operator and the PC pump supplier. Thus, it is very critical to select a supplier who has vast knowledge of the pump and its' applications. This subject will be discussed further in Chapter 4.

If an application is properly screened and the PC pump is sized correctly, an operator can expect a high level of success. Throughout the history of the PC pump as a method of artificial lift, it has been operated in applications where other methods were unable to pump, resulting in poor run times and poor reputations in many cases. On the other hand, in these "problem wells," the PC pump had a longer run time than other lift methods, in most cases.

In applications that pose no unusual problems such as solids, gas locking, etc., the PC pump has been successful in achieving long run times.

The key to reliability resulting in long run times is the proper rotor/stator fit. You should strive to have a rotor/stator compression fit that gives the lowest torques possible while maintaining high volumetric efficiencies.

Application Flexibility

Under the current conditions of the oil and gas industry, the end users need products that are efficient and flexible to help keep their lifting costs to a minimum.

This is an area where the progressing cavity pump will out perform other lift methods. The PC pump can be utilized in a large variety of applications such as light oils up to 35°–38° API oils, heavy oils as low as 8° API, water source wells, dewatering coalbed wells, or other gas wells. The PC pump can be utilized in applications where gas locking occurs with other methods of lift, where solids create problems, and where low profile units are a necessity.

The progressing cavity pump has excellent flexibility in regard to production changes. The pump speed (RPMs) can be altered to match production rates. This can be accomplished by changing sheaves, utilizing hydraulic drives, or utilizing variable frequency drives.

Automation has been developed for the variable frequency drives, which make it possible for the drive to change speeds automatically as the well conditions change, thus protecting the stator from a run dry operation. We will discuss this topic in more detail in Chapter 7.

Environmental Benefits

The progressing cavity pump has many advantages over traditional methods of artificial lift when it comes to environmental concerns. The greatest advantage is the lack of obtrusiveness. The PC pump offers low-profile surface equipment, making it conducive to the environment. This is a big benefit in areas where farmers are utilizing irrigation systems and in populated areas.

In regard to populated areas, the PC pump has advantages over beam pumping units not only in its smaller size but also in the fewer number of moving parts on the surface, thus limiting exposure to injury.

Another area of benefit is in reducing noise levels. The PC pump performs excellently in regard to noise levels; generally, the prime mover is the only noise heard. Some PC pump suppliers use soundproofing installations in their belt guards to reduce the noise level. Also, in some areas such as northern Michigan, suppliers have soundproofed the hydraulic drive units and installed the drives in buildings.

In most cases, leakage of the packing area on the PC pump drive unit is kept to a minimum. However, it is difficult to have a "no leak" condition on a rotating shaft. Some drive manufacturers do a better job of this than others.

CHAPTER 3

THE LIMITATIONS AND DISADVANTAGES OF THE PROGRESSING CAVITY PUMP AS AN ARTIFICIAL LIFT METHOD

Depth and Volume Limitations

The Progressing Cavity Pump continues to evolve to meet the wide variety of oil and gas applications the industry has to offer; however, the pump does have its limitations.

The biggest limitation on depth and volume is related to the torsional strength of the sucker rod string. There are no specifications provided by rod manufacturers or required by API on torsional strength of sucker rods. Sucker rods are only distinguished by tensile strength values. This is one area in which there is a need for research and development.

The other area of depth and volume limitation is in the manufacturing process. There is a limit to the physical capabilities of the machining lengths. The equipment used to machine the rotor and stator have a maximum length capacity of just over 200 in., which results in a depth capability of 4,000 ft. in a single rotor and stator combination.

There are also limits to the lengths in which the elastomer can be successfully injected into the stator during the molding process.

The manufacturers and/or the suppliers reach greater depths by one of two methods. One method is referred to as "phasing." In this method, two pumps of the same size and design are matched up and welded together, resulting in more pump stages or pressure capability.

The other method is referred to as a "tandem." In this method, two pumps of like size and design are run together in series. The pumps are connected together by a short sucker rod (a pony sub usually 8 to 10 ft. in length) for the rotor and a short tubing sub of similar length for the stator, once again resulting in more stages or pressure capabilities without the welding.

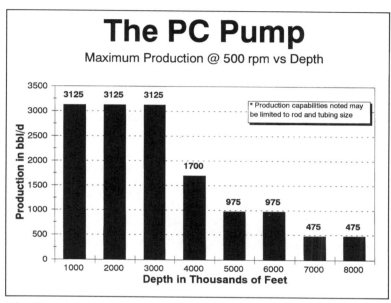

Fig. 3-1

One advantage of running pumps in tandem over phasing is the stocking and handling of the pumps, since it is much easier to handle a short piece than a longer one. However, if the end user has selected a quality supplier, the handling of phased pumps will not be a concern.

The graph in Figure 3-1 shows the current depth and production capabilities of the Progressing Cavity Pump.

TEMPERATURE LIMITATIONS

The temperature range of the Progressing Cavity Pump is directly restricted by one component: the stator elastomer. Manufacturer's of the PC pumps and their elastomer suppliers need to place more emphasis on this component.

The temperature ranges of the three common elastomers that are on the market are as shown:

Medium Acrylonitrile	Ultra-High Acrylonitrile	Highly Saturated Hydrogenated Nitrile
180°F	200°F	275°F

Refer to the discussion in Chapter 1 on the materials of construction for more detail on the elastomer characteristics.

The effect that temperature has on the stator elastomer is ultimately one of efficiency. As the elastomer reaches its maximum temperature range, thermal expansion takes place, thus resulting in a decreased cavity capacity and an increase in rotor/stator friction (higher torque). This is yet another strong case for pretesting the rotor and stator in the same parameters as are expected in the well.

If the elastomer is exposed to temperatures higher than its rateds, there will be a catastrophic failure, with the elastomer coming out of the stator tube.

Many manufacturers continue research and development in the area of elastomers. The marketplace will benefit from these efforts in the near future.

APPLICATION FLEXIBILITY LIMITATIONS

One obvious limitation of the Progressing Cavity Pump is its depth. However, there are some other areas in which the pump is limited.

An area of concern is high temperatures. Since all of the elastomers have a maximum temperature range, it is important to know to what temperature the pump will be exposed.

Progressing Cavity Pumps have been tried in both steam floods and huff-and-puff operations over the years with very limited success.

The temperature ranges in most steam floods exceed the temperature range of all the pump elastomers. The pumps, however, are a good choice in applications that are on the fringe of a steam flood due to lower temperatures and the PC pump's ability to pump low-gravity oil with solids.

The problem with utilizing a PC Pump in a huff-and-puff operation is selecting the proper rotor/stator fit.

In the past, attempts were made to start out with a loose fitting rotor/stator in anticipation of the fit getting tighter as temperature is increased. However, the problem that occurs is that the initial fit is so loose that the volumetric efficiency suffers. Or in some cases, the fit may be alright initially, but when the temperature is increased, the rotor/stator fit becomes so great that torque becomes a problem.

Thus, the true problem is changing temperature, with the only solution being to change the rotor when higher temperatures are expected, which is not practical. It is wise not to utilize Progressing Cavity Pumps in this application until the manufacturers develop elastomers that can better handle high temperatures.

Another area of application limitation is in the pumping of high-gravity crude oil. As discussed earlier, the maximum gravity oil that most PC pump elastomers currently on the market can handle is 35° API to 38° API. The effect of high gravity on the elastomer results in swelling due to the aromatic hydrocarbons in the fluid. It is strongly suggested that the operator choose a PC pump supplier who will take fluid samples and have the well fluid checked for compatibility with the pump elastomers to avoid any potential problems.

Sour oil and/or H_2S application is another area in which the Progressing Cavity Pump is limited. The PC pump has been tried in H_2S wells over the years, again with very limited success. There have not been a large number of PC pumps applied in this type of application; of the ones that were installed, run times were shorter than the operators would have liked.

The effect of H_2S on the pump elastomer results in a hardening of the outer layer of the elastomer. The H_2S apparently penetrates the elastomer, making it hard and brittle. Some manufacturers and pump suppliers claim successful run times in applications of 30,000 ppm H_2S with their hydrogenated nitrile elastomers. It does appear that this compound receives the best results. From what I have experienced and was able to learn about these applications, the lack of a successful run time may not be due to just one component, H_2S, but to a combination. In many of the applications where the pumps were run in H_2S, the well fluid being pumped also had high-gravity oil. Thus, a combination of H_2S and high aromatics might be causing the problems. More research is needed before these applications can be successfully pumped with a Progressing Cavity Pump.

Other applications in which the Progressing Cavity Pump may have limitations are in well treating programs. An operator again should select a manufacturer or pump supplier who will test the operator's treating chemicals for compatibility. Some chemicals have a tendency to cause the elastomer to swell.

On the other hand, by utilizing a Progressing Cavity Pump, the operator may find that the need for his treatment program is no longer necessary. Applications where this may be true are low-gravity oil applications where the operator is hot oiling the well or treating with chemicals to reduce the viscosity of the oil being pumped. It should be noted that, due to the current elastomer maximum temperature range, hot oiling is noneffective. Most hot oiling treatments are conducted at a temperature greater than the maximum range of the elastomers.

Another application in which the treatment may be discontinued by utilizing PC Pumps concerns paraffin. Due to the constant nonpulsating flow of the fluid and the rotation of the rod string, the effect of paraffin build-up is greatly slowed down, which is a benefit because most chemicals used for paraffin treatment cause swelling of the elastomer to some degree.

The Progressing Cavity Pump is no different than any other method of artificial lift equipment in regard to its applications, meaning that it does have limitations and there are some applications where caution is needed.

Service and Experience Limitations

End users need to investigate at length service and experience limitations before selecting a pump supplier. Currently, there are no Progressing Cavity Pump manufacturers that market directly. All of the manufacturers market through some sort of sales and service group. Even if the manufacturers marketed directly, it would not guarantee that the end user would receive quality service. This is due to the fact that few PCP manufacturers have true experience in the oil and gas industry in applying the product.

The quality of service and experience is as varied as the suppliers themselves. It has been my experience that the larger pump suppliers who have multiple product lines lack the expertise and experience to supply quality service. In contrast, the smaller suppliers who sell only Progressing Cavity Pumps or have a division solely dedicated to the PC pump tend to have more experience and provide a higher level of quality service.

You want to look for a supplier who has been associated with the Progressing Cavity Pump for a good number of years so you can benefit from their experience. Also, make sure that the supplier you select is dedicated to service via field technicians, service trucks, etc.

I would suggest that, when talking to the pump suppliers about an application, the end user ask the supplier about his philosophy of applying the PC pump to downhole pumping applications. Also, the end user should note the degree of thoroughness used in screening the application. Asking for well data, production history, and fluid analysis, and taking fluid samples indicate that the supplier has experience in screening applications.

Another area of service that the suppliers should offer is pump testing. The end user should ask the supplier if he has pump test facilities in the market area. This will give the user some level of comfort to know that if he experiences a pump failure, he can have the pump performance tested.

The key to a successful Progressing Cavity Pump installation is having a supplier with the expertise and knowledge to appropriately apply the pump to a given application while offering quality field service support.

SALVAGE VALUE LIMITATIONS

The overall Progressing Cavity Pump salvage value is fair. There is not a large demand on the open market.

The equipment, however, is easily moved and drive heads can be inspected easily and reused, whereas the rotor and stator need to be performance tested before reusing them in a well. An operator should want the pump performance tested before spending the money on a service unit to install the pump.

You want to be very careful in purchasing used stators since its wear is internal and usually not visible to the eye. Additionally, the stator's elastomer can have swell present or gas permeation (gas bubbles), resulting in high torque problems if the stator could operate at all.

The rotors can be easily inspected and, in fact, can be rechromed if wear is present as long as the base metal is not damaged. You will want to contact the respective manufacturer to get the specifications on rotor dimensions.

Some Progressing Cavity Pump suppliers will take used equipment in exchange for other PCP equipment. Operators who have obsolete equipment could take advantage of this.

CHAPTER 4

THE PROGRESSING CAVITY PUMP DOWN HOLE APPLICATIONS

Due to the flexibility of the Progressing Cavity Pump, it has been used in a wide variety of oil and gas applications. The PC pump has been utilized not only in lifting fluids from oil and gas wells but also in water source wells, environmental test wells, caisson pumping applications, and as an injection pump. All of these applications will be discussed in this chapter as well as the proper procedure for screening and sizing applications.

COMMON PC APPLICATIONS

With the decline in activity in new well starts in the United States, the majority of new well applications that have utilized the Progressing Cavity Pump have been for the dewatering of both conventional gas wells and coalbed methane wells.

Over the last five years, there have been over 1,000 PC pumps installed in dewatering applications. Much of this activity was directly related to a special tax credit for unconventional fuel sources. The major concentration of PC pumps in this type of application have been in the Black Warrior Basin in Alabama, the Antrim formation in northern Michigan, the San Juan Basin in southern Colorado and northern New Mexico with a lesser amount in the Raton Basin in northeast New Mexico, and in the Uinta Basin in Utah.

The Progressing Cavity Pump is also being utilized in water floods. There are big advantages in using PC pumps in this type of application, the most obvious to the end user is the savings in capital costs. Operators have been taking advantage of lower capital costs on wells in which the pumping unit no longer is adequate to move the volume of fluid that is desired. Thus, operators turn to the PC pump instead of buying a larger beam unit or an electric submersible pump at a higher cost.

One should also look deeper at the potential savings in annual lifting costs by utilizing smaller electric motors. The PC pump, with its higher volumetric efficiencies and rod string that is lifted, enables you to use smaller motors to perform the same amount of work as the beam pumping unit or electric submersible pumps. Examples of this benefit were given in Chapter 2.

One of the most common Progressing Cavity Pump applications worldwide is in heavy oil applications. The PC pump has been utilized for fifteen years in this type of application in Canada where it has become the pump of choice.

One advantage in using a PC pump in this application is that you can eliminate rod drop problems that beam units experience as well as the need for sinker bars. Another advantage in this application is that the PC pump can handle the solids associated with the heavy oil better than the conventional down hole pumps.

Another application that is becoming more common for the PC pump is pumping oil and gas wells where irrigation systems are being used by the farming industry. The PC pump is the preferred method of lift due to its low profile and its capital cost when compared to a low-profile beam pumping unit. The PC pump is being used in this manner in the Hugoton gas field in southwest Kansas. In this application, it is vital to have an automated pump controller due to the low volume of fluid being produced. Progressing Cavity Pumps have been tried in years past in this application without a controller and have failed due to run dry operations.

Another factor in the failure rate in this application was the method of operation. Most operators had their PC pumps operating at high speeds and on a time cycle that made it difficult to know when the well was pumped off. Additionally, the fact that it is more efficient to operate at slow speeds on a 24-hour basis, maintained a constant draw on the formation as opposed to a cyclical one.

UNUSUAL PC APPLICATIONS

With the long proven history of the Progressing Cavity Pump in both the industrial markets and oil and gas markets, it has been used in a great number of unusual applications.

Since the Progressing Cavity Pump proved its versatility in conventional oil and gas wells, it was simply a matter of time before operators found other applications in which they could benefit from the PC pump's flexibility. One such application is referred to as caisson pumping.

There have been a small number of caisson pumping applications installed on offshore platforms in the North Sea and also in the United States. In the

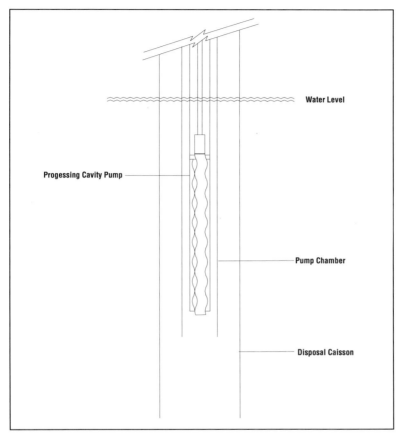

Fig. 4-1

future, there will surely be a bigger marketplace for this type of application than what is currently being used. However, there are few, if any, pump suppliers working this market especially in the United States.

In a caisson application, a PC pump is installed in a large pipe that is used for disposal on an offshore drilling platform to retrieve a variety of fluids. A schematic diagram of the pump assembly is presented in Figure 4-1. The stator is run in on a tubing string inside the pump chamber or casing, usually to a depth of 60 ft.

The rotor is installed inside the stator on the sucker rod string. The fluids are pumped through the PC pump up the tubing string into a flowline to a sump tank.

The fluids being pumped consist of sea water, production water, diesel oil, crude oil, and a variety of chemicals. Also, a variety of solids can be present; however, most settle and fall out of the caisson pipe.

The Progressing Cavity Pump is utilized in caisson pumping because of its versatility in handling different types of fluids and its compact, low-profile design.

All the other Progressing Cavity Pump benefits apply here also: capital cost, power consumption, and low noise levels.

This brings up the subject of other applications on producing wells offshore. Again, there have been very few operators taking advantage of the PC pump in this type of application.

There are many benefits in using a PC pump on a production platform. The compact size and low noise level provide the greatest benefits, along with the cost savings.

For the most part, the PC pumps that have been utilized to date on production platforms have been in shallow applications, 2,000 ft. or less, with low production volumes of under 300 BFPD.

With all of the advancements of the Progressing Cavity Pump product line, I believe the pump could be applied to a larger number of offshore applications.

Some of the reluctance of operators to utilize the PC pumps has to do with the lack of knowledgeable pump suppliers in their market areas. Also, a great number of offshore applications are deviated wells, which require more expertise in installing pumping equipment. Deviated wells will be discussed later. Operators of offshore production may want to take a second look at the Progressing Cavity Pump system for their application.

Another unusual but somewhat common application for PC pumps is in slant wells. This type of production is popular in Canada, mostly in heavy oil applications but it is also used in some light oil applications. This method is used when the operator starts out drilling the well from the surface at a 30° slant then continues deviation to the pay zone. The purpose of entering the zone at a slant is to expose more area of the pay, much like a horizontal well.

The Progressing Cavity Pump is utilized because of its compact drive head that mounts directly to the tubing string via the pumping tee. Other advantages to this method are that it is common to have up to ten wells drilled from one location, thus reducing the surface damage, having one lease road to a well vs. ten, and all of the equipment can be serviced from one location.

When utilizing a Progressing Cavity Pump in this type of application, the operator must use rod centralizers or rod guides to eliminate rod and tubing wear.

The PC pump has been utilized in this horizontal well applications but not in the same volume as in slant wells. Again, when installing a PC pump in a horizontal well it is necessary to have all available data on the deviation before running the pump in the well. It is also important to utilize rod centralizers or rod guides in this application.

The Progressing Cavity Pump has been installed in dual completion wells both in Canada and in the United States.

These applications are no different than the dual completions in which beam pumping units are utilized. The advantage of using a PC pump in place of a conventional method is the size of surface equipment. The Progressing Cavity Pump's compact surface drive heads, which mount directly onto the well head via a pumping tee, reduce the surface area needed and eliminate the need for skids or pads that are required for a beam pumping unit. The other big benefits to the operator are in the savings on capital cost and power consumption.

The Progressing Cavity Pump is also used in ground water testing. This is not an oilfield application; however, the pumps are installed in wells similar to an oilfield application. The purpose is to test fresh ground water for environmental concerns. The pumps can be used around gasoline stations or chemical plants to make sure the ground water has not been contaminated.

In these applications, a different stator elastomer is used. Typically, a fluoroelastomer is utilized because of its excellent resistance to solvent and acids. Also, a smaller staged pump is needed (usually a 1- to 2-stage pump) due to the shallow depth of the test wells.

One unusual application for the Progressing Cavity Pump could be to provide the end user with a number of benefits as a downhole disposal pump.

This method was considered a few years ago to limit the risk of environmental exposure and to reduce the need of pumping facilities and flow lines.

The PC pump would be utilized in a well that produces water and gas from an upper formation and was drilled through a lower formation that would be used for disposal.

The type of PC pump that is used for this purpose

is referred to as an industrial PC pump. It has a lead opposite to that of an oilfield PC pump, resulting in the ability to pump fluid down with a clockwise rotation.

The pump would be set between the producing zone and the disposal zone, utilizing a packer to isolate and seal off the two zones. A perforated tubing sub would be used above the packer to allow fluid to enter the tubing string and the intake of the stator, which would be set below the packer.

However, a few areas of concern would need to be addressed. First, the wear on the rod string and tubing resulting from the rods rotating in the tubing with no fluid.

Rod guides and coated tubing would have to be utilized to reduce the wear. The other area of concern would be knowing if the pump had adequate fluid entering the intake of the pump to avoid a run dry operation. Some sort of pump off controller would be needed. There have been some technological advancements made recently in this area that could remedy this problem.

This same design was tried once before in a close loop injection system, where the PC pump was set between the water source zone and the formation that was being water flooded (producing zone). During this test, pressure gauges were utilized to help monitor pump pressure and performance.

This field test was successful in proving that a PC pump could be utilized as a water flood injection system.

As the PC pump continues to evolve technically and the experience level increases, it will be used in an even greater variety of applications.

SCREENING AND SIZING PC APPLICATIONS

The most critical element in assuring a successful Progressing Cavity Pump installation is the screening of the well conditions beforehand and selecting the correct pump size for the application.

Most of the pump suppliers provide the end user with an application data sheet in order to screen the application. (See Figure 4-2 for an example of this form.)

The information that is requested on the form should be easily attainable for most producers. The purpose of gathering this information is to look for any red flags or warning signs that might eliminate using a Progressing Cavity Pump or that require special consideration.

Examples of such warning signs include something as simple as the required flow rate or the depth of the application being out of the range of a PC pump. Another case that could eliminate the use of a

PC pump is an application with fluids that are not compatible with the pump's materials of construction.

Examples of applications that would require special consideration are those without electrical power, in which case the end user would have to consider natural gas engines or natural gas hydraulic units. Applications with H_2S would require special elastomer consideration as would high-gravity oils (30° API to 38° API).

It is vital when considering installing a Progressing Cavity Pump in a new field or formation where PC pumps have not been utilized, that the pump supplier get a fluid sample from the well and have it tested for compatibility with the stator elastomers.

As with any pump, the more information you obtain about the materials being pumped, the better chance you have for a successful run time.

Sizing a Progressing Cavity Pump for an application is very simple. There are two key factors to consider: the total dynamic head and the required flow rate. The total dynamic head is the dynamic fluid level and the flow line pressure converted over to feet combined.

After determining the abrasive characteristics of the fluid being pumped (percent of solids in solution), you will need to refer to the manufacturers speed guidelines for abrasive fluids. Most manufacturer's recommend avoiding operating the pump in the upper half of its speed range. In most cases, they recommend installing a larger capacity pump to achieve the desired production rate at a lower speed and one with more stages, resulting in lower pressure per stage, which will extend pump life in an abrasive environment.

Upon knowing the desired production rate, the total dynamic head, and any special considerations such as abrasive fluids, high temperatures resulting in elastomer expansion, or high aromatics resulting in possible swelling of the elastomer, you are able to select a pump model that is best suited for the application.

As discussed earlier, by selecting a pump supplier who "matches" rotors and stators to a given set of well conditions, he is able to supply the end user with a pump that can achieve high volumetric efficiencies under any well conditions.

After selecting a pump for an application, the surface drive head needs to be selected. The important factor is sizing a drive head with a bearing capacity large enough for the application, resulting in long LIO life. You can refer to the manufacturers for LIO ratings on their drive heads.

It is also necessary to size the proper prime mover to operate the pump as well as to select accessory equipment such as rod guides, anchors, separators, and flow controls, all of which will be discussed in Chapter 7.

PROGRESSING CAVITY PUMP APPLICATION DATA SHEET

COMPANY: _____
WELL LOCATION: _____
FIELD: _____

DATE: _____
REPRESENTATIVE: _____
PHONE: _____
FAX: _____

GENERAL WELL DATA

Total Depth: _____
Perf Depth: _____
Formation: _____
Casing Size: _____
Tubing Size: _____
Rod Size: _____ Grade: _____
Power Source: _____

PRODUCTION DATA

Current Production Rate: _____
Required Rate: Min. _____ Max. _____
Statis Bottomhole Pressure: _____
Dynamic Fluid Level: _____
Water Cut: _____
GOR: _____
Bottomhole Temp: _____
Flowline Back Pressure: _____

FLUID DATA

Oil API: _____
Oil Viscosity: _____
Specific Gravity Water: _____
Solids (sand): _____
Wax: _____
Treating Chemicals: Mfg. _____ Brand Name: _____

H^2S: _____ (wellhead concentration)
CO_2: _____ (%)
Aromatics: _____
(volume/characteristics)

ADDITIONAL COMMENTS:
_____
_____
_____

Fig. 4-2

SELECTING A PC PUMP SUPPLIER

Reference has been made throughout this book to the importance of selecting a Progressing Cavity

Pump supplier. This decision could hold the key to a successful PC pump installation.

Bear in mind that none of the Progressing Cavity Pump manufacturers are currently marketing their product directly to the oil and gas producers, which means that the end user has to rely on the sales and service of the manufacturer's distributors.

The greatest barrier to the expansion of the Progressing Cavity Pump market has been the lack of knowledge and expertise in the industry, which also results in a high failure rate.

If you take a look at the short history that the Progressing Cavity Pump has had as a method of artificial lift in the oil and gas industry, you will find that a great number of pump suppliers have come and gone in the marketplace, leaving the end user without adequate field service. And the manufacturer has not supplied the industry with adequate field service expertise. This fact may account for the number of failed distributors and most certainly accounts for the high pump failure rate.

Since the manufacturers are not marketing the product directly to the end user, the responsibility of field service expertise falls back on the distributors, many of whom do not have an understanding of the product or how to properly use it in given applications.

Each manufacturer is responsible for training the distributors. The problem, though, is that while the manufacturer has the knowledge and expertise to design and build the progressing cavity pump, few have the knowledge and expertise to apply the product to the application. This makes it difficult to properly train the distributors to service the product.

Thus, it is vital that the end user select a pump supplier with a proven track record. It is suggested that you get a reference list of the supplier's PC pump users and contact them to see if they are happy with the product quality and the level of service that they are receiving.

It is also strongly recommended that the producer select a pump supplier who not only provides a "new pump performance test" but will also "match" the rotors and stators to a given set of well conditions. The pump supplier should provide the end user with the "fitted" pump and with the test results that demonstrate how the pump will perform in his application.

In addition, the end user will need to discuss with the pump supplier what their philosophy is on PC pumps, i.e., matched pump elements, operating speeds, providing performance data, etc. The pump supplier should have the ability to retest pumps in a timely manner. The end user should keep in mind, however, that it is not practical or economical for the pump supplier to have a test facility in each marketplace, so a good supplier will have adequate inventory on hand to develop a reserve pump for the end user. The supplier should have the expertise to retest a pump and then instruct the end user on what type of well conditions that particular pump will need to operate in to achieve the greatest run times.

The end user will also need a pump supplier who has qualified, experienced service technicians to assist in the installation and maintenance of the product along with the ability to conduct pump performance analysis for the end user. Thus, it is important that the end user select a supplier whose service technicians have been equipped with the proper tools and vehicles to perform an efficient job.

You will find that not all of the Progressing Cavity Pump suppliers meet this requirement. Some suppliers have field personnel who are equipped with a pickup truck and some hand tools. Others are dedicated to the long-term market support of the PC pump and have technicians equipped with service trucks with hydraulic booms mounted on them and fitted with all the tools that are needed to perform service work. In some cases, the pump suppliers have larger service trucks that not only have the hydraulic boom but also are equipped with compressors to power air tools and generators for electric tools and welders. In the latter case, the supplier is able to offer the end user "complete service." This supplier will be capable of working on pump systems with a much higher efficiency and a lower risk of personal injury. This type of supplier is able to perform a great number of services on a pump system without the end user having to call out a service rig or other boom trucks to perform a task. The end result is a lower operating cost to the end user.

Photos of such a service vehicle are shown on the following page.

Another service that the end user should ask if his pump supplier can provide is the capability to perform power consumption studies.

The end users need to think in terms of overall operating cost not just capital cost. A good pump supplier will have the equipment to monitor kilowatt usage to show the end user just how much it is costing him to produce his well. He is also able to supply the end user with power consumption data and other performance data in a monthly report through a Peak Performance Program.

With the current state of the oil and gas industry, in which many operators do not have the human resources and time to gather this type of data, a pump supplier with this level of expertise and capability is a welcome part of their team. This type of pump supplier is able to perform the data gathering on all types of lift methods as well.

Thus, you are looking for a pump supplier who through the quality products and service support they provide has your operating costs in mind. By utilizing this level of service, you will discover that you can lower your operating costs via lower power consumption and shorter amounts of down time due to improved mechanical and volumetric efficiency, i.e., fitted pumps, with the end result of long, efficient run times.

Thus, it is vital that the end users do their homework in selecting a PC pump supplier in order to benefit from the new technology that is available in the marketplace.

CHAPTER 5

INSTALLING THE PROGRESSING CAVITY PUMP DOWN HOLE

DETERMINING PUMP SETTING DEPTH

A number of factors that can play a part in determining the pump setting depth of a Progressing Cavity Pump are: the dynamic fluid level, solids that are present, gas production, and perforation depths in relation to total depth, among others.

For the most part, it is recommended to set the pump below the perforations or pay zone in both oil and gas applications. But it is especially important in dewatering of gas applications such as those found in coal seams in the San Juan Basin, Black Warrior Basin, and Raton Basin, Appalachian Basin, Uinta Basin.

In the dewatering of gas wells, the practice of setting the pump below the pay zone or perforations is to allow the free gas to break out and move up the annulus rather than passing through the pump.

This practice is utilized to maintain the pump's volumetric efficiencies. A Progressing Cavity Pump will not gas lock like a conventional reciprocating pump; however, gas passing through the pump will have an affect on the pump's efficiency.

In some dewatering applications it is not possible to set the pump below the pay zone or perforations for a variety of reasons. In these cases, it is normal practice to set the intake of the pump in the middle of the perforations or to run a tail pipe below the stator. In applications where solids are not a problem, a conventional gas/mud anchor can be used below the stator of the pump. However, if the application calls for a large volume of fluid to be produced, the gas/mud anchor may be ineffective due to the velocity of the fluid.

In some applications, a producer may decide to produce a well where the perforations are deeper than the capability of the Progressing Cavity Pump, so he may elect to set the pump up hole and maintain a high fluid level. For the most part, this is done in applications that have a strong water drive or in a water flood where there is a low risk of pumping off.

Other applications in which the producer may elect or be forced to set the pump above the perforations are in wells that have no rat hole, where solids are a concern, and/or where clearance is a problem.

INSTALLATION OF THE STATOR AND ROTOR

The installation of a Progressing Cavity Pump is fairly simple and requires no additional equipment that is not common to conventional pumps.

A good pump supplier will assure that all the proper connections and fittings are on location at the time of installation.

The stator is attached to the first joint of tubing. You will need to refer to the manufacturer's pump specifications for the stator thread type and size.

Typically the stator is threaded on both ends so accessories such as gas/mud anchors or tubing anchors can be utilized where necessary.

It is important that the tubing string being used is large enough to accommodate the expected flow rate. If it is necessary to swage down to smaller tubing, it is recommended that you change over at least six feet (6') above the stator.

When installing the stator and tubing, power tongs should be used and the tubing should be tightened to API specifications to insure that the tubing does not back off if torque is present during operation.

TABLE 5-1

	J55 EUE Tubing	
	Optimum Tq.	Max Tq.
Tubing O.D.	(ft-bl)	(ft-bl)
2 3/8 EUE	1,290	1,610
2 7/8 EUE	1,650	2,060
3 1/2 EUE	2,280	2,850

	C75 EUE Tubing	
	Optimum Tq.	Max Tq.
Tubing O.D.	(ft-bl)	(ft-bl)
2 3/8 EUE	1,700	2,130
2 7/8 EUE	2,170	2,710
3 1/2 EUE	3,010	3,760

Attach the stator to the first joint of tubing that is specified for the pump model being used. Make sure the tag bar is on the bottom of the stator. All accessory equipment should already be attached to the bottom (intake) of the stator. Tighten all connections to the API specifications (see Table 5-1).

It is recommended that you measure the stator as well as all the tubing and tubing subs being used. Begin by measuring the stator from the top (discharge) to the stop pin or tag bar on the bottom (intake) of the stator. Tally and record all the measurements as the tubing is run into the well.

As discussed previously, set the stator below the perforations wherever possible.

By recording your tubing tally you can determine the number of sucker rods necessary to land the rotor into the stator. The pump supplier will have a selection of pony subs on location to be used for spacing of the pump.

The rotor is run into the well on the end of the sucker rod string.

Attach the rotor to the first full length sucker rod. Do not install a pony rod next to the rotor. Also, do not run in any sucker rods that are bent, corroded, or damaged. Snap tight all rod connections during installation.

The last few sucker rods to be run in should be installed slowly. Before entering the rotor into the stator the rod string weight should be recorded if a weight indicator is being utilized.

Before taking any spacing measurements, all flow tees, hammer unions, or flanges should be attached to the tubing at this time.

When inserting the rotor into the stator, the rod string may rotate to the right. At this time, a decrease in the rod weight may be noted on the weight indicator.

Put all the rod weight on the tag bar/stop pin.

Make a mark on the sucker rod at the flow tee or hammer union, whichever is being utilized. This mark will be at zero rod weight. You may want to recheck your landing by pulling up and setting back down on the tag bar/stop pin again. The zero rod weight mark should still be at the flow tee or hammer union. When the rod weight is resting on the tag bar/stop pin, slump can occur.

Pick up the rod string off the tag bar/stop pin until rod string weight is achieved, then refer to the weight indicator. The slump will be taken out and the rods will be in tension.

Mark the sucker rod at the flow tee or hammer union again; this is the rod string weight mark.

You will need to get the required rod stretch recommendation for the rod size and pump model being utilized from the respective manufacturer or pump supplier.

Pick up the rod string the required distance and mark the rod at this point—this becomes the operating point.

If you are using a drive head that has a hollow shaft (which is the most popular head), you will need to add the length of the drive to the operating point. This becomes the clamping point.

You will need to pull enough sucker rods out of

the well to allow for the polish rod to be installed in its place.

Measure the distance from the operating point to the next coupling on the rod string below the operating point. This measurement plus the length of the drive head become the amount of polish rod and pony subs (if needed) that need to be reinstalled in the well. NOTE: Allow for 6 inches (24 inches maximum) of polish rod to protrude above the polish rod clamp.

The next step is to install the drivehead.

DRIVE ASSEMBLY INSTALLATION

There are a variety of drivehead assemblies in the marketplace; however, the most common one today is a hollow shaft design that utilizes a polish rod and polish rod clamp. You may need to refer to each respective manufacturer for installation instructions.

The hollow shaft design allows the polish rod to run through its bearing housing as well as its stuffing box assembly.

An advantage of this design over a solid shaft design is that there is ease in spacing the pump and making adjustments without having to take the drivehead off the wellhead. With the solid shaft design, you would have to take the drivehead off the wellhead to change spacing, opening the well up to the atmosphere. If you experience a bearing failure or brake failure on the hollow design, you are able to clamp off the polish rod in the stuffing box area and unbolt the bearing housing, leaving the well protected from the atmosphere while changing the bearings or the brake.

This brings up another important point. You should select a drivehead that incorporates a braking device for safety. Some Progressing Cavity Pumps will motor backward when shut off and the braking device insures that the "backspin" is controlled. This subject will be discussed further in Chapter 6.

To continue the installation of the drivehead, you will need to install any pony rods necessary to achieve the required "pull back" or "rod spacing." Remember to take the drivehead height into account when installing the polish rod; the difference in overall length can be made up with the pony subs.

With the pony rods installed in the well, put a 2 ft. pony rod on the top of the polish rod and attach it to the rod string. Lower the polish rod into the well, leaving enough exposed to mount the drivehead over it. At this time, you need to place a clamp on the polish rod resting on the flow tee or hammer union.

Remove the rod elevator and 2 ft. pony rod. Lubricate the polish rod before lowering the drivehead over the polish rod.

Pick up the drivehead with the service rig's cat line or the pump supplier's boom. Be sure to keep the drivehead balanced while you are lowering it over the polish rod so as not to bend the polish rod.

After you have lowered the drivehead onto the polish rod, place the 2 ft. pony rod and rod elevator back onto the polish rod. Remove the clamp and continue lowering the drivehead onto the flow tee or hammer union.

Thread the drivehead into the flow tee or hammer union making sure not to overtighten.

Once the drivehead is mounted, you can place the polish rod clamp on the top of the polish rod at the clamping point and tighten. Now you are able to lower the polish rod down the rest of the way until the clamp is resting on the drivehead. NOTE: Again, it is important to have a minimum of 6 in./maximum of 24 in. of polish rod sticking up above the drive clamp. Also, you should remove the pony rod from the top of the polish rod, leaving just the rod box.

A schematic diagram of the progressing cavity pump assembly and drive assembly is presented in Figure 5-1.

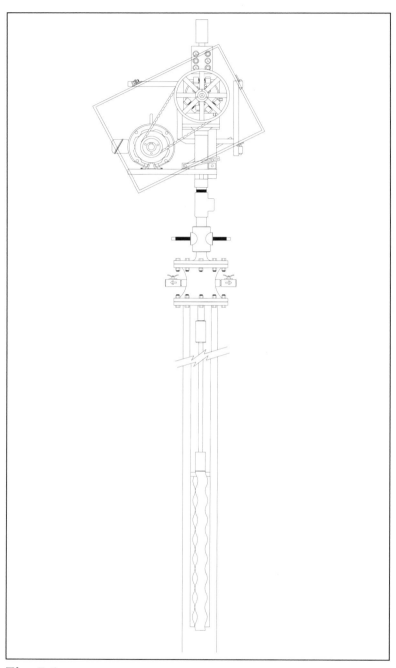

Fig. 5-1

Pre-Start Check List

It is important to go over a pre-start check list prior to operating your Progressing Cavity Pump system.

First, it is a good practice not to release the service unit until you are comfortable that your PC pump is operating correctly. If the pump was spaced out incorrectly and the rotor was on the stop pin/tag bar you would need the service unit to re-space the pump.

Before starting your PC pump, you will need to check the prime mover rotation to assure that the rod string will rotate clockwise. After checking rotation, be sure to check the alignment of the belts and sheaves that are being utilized, if any. You will also want to be sure that the belt guard is in place.

Before startup, you will want to see to it that the bearings, brake, and packing are lubricated properly. The packing should not be tightened down until the well has pumped up.

You will also want to double check to make sure that all valves are open on the flow line before operating the pump.

It is a good idea to have a fluid level reading on the well prior to and after startup.

Low profile pump application with irrigation system.

Soundproof building for hydraulic drive unit.

Common PC pump application

Common PC pump application

Common PC pump application

Size comparison of PC pump to beam-balanced unit.

Unusual PC Pump Applications

The PC pump as an offshore production tool.

Slant well applications

A multi-well lease using PC pumps reduces lease surface damage.

The PC pump in a dual-completion well application.

The supplier dedicated to long-term support of the PC pump will use service trucks equipped with mounted hydraulic booms and the tools and equipment for "complete service."

Variable frequency drive unit.

Cross-section views of the PC pump drive assembly.

Hydraulic drive unit.

Anti-friction unit

CHAPTER 6

OPERATION AND MAINTENANCE OF THE PUMPING SYSTEM

One of the benefits of the Progressing Cavity Pump is its ease of operation and the low maintenance that is required.

The PC Pump, however, is like any other piece of equipment—it will require some attention to assure that the end user is able to receive maximum life out of it.

For the most part, the only maintenance that is required consists of oil changes and the greasing of packing on the manufacturer's recommended time frame.

PC PUMP OPERATING PERFORMANCE

After your Progressing Cavity Pump has been installed, you should experience trouble-free operation.

You will want to check the pump's performance occasionally to assure yourself that it is performing to the manufacturer's specifications.

This can be achieved if, as discussed in earlier chapters, the end user has selected a pump supplier who not only offers performance testing of new pumps, but one who also offers "in-well performance analysis". This is a service that is provided by few suppliers, but can be very beneficial to the end user.

The pump supplier will perform tests and take readings on the pump and then supply the test data results to let the end user know how his PC Pump is performing. The type of information that can be supplied includes the volumetric efficiency, dynamic fluid level, pump speed (RPMs), flowline and casing pressure, and kilowatt usage of the electric motor.

The pump supplier can also provide performance and production curves on the pump that is in operation, similar to the performance data supplied on a new pump.

A good supplier will also offer a maintenance program to assure that the producer's equipment is being maintained according to the manufacturer's specifications.

As far as maintenance of the PC Pump is concerned, there is basically no maintenance during operation, other than checking the well's productivity. However, prior to installation and at any time the pump is pulled out of the well, it should be inspected and performance tested if needed.

The rotor should be inspected for wear from normal operation or abrasion. The bottom of the rotor should be checked for wear to make sure that it has not been riding on the tag bar/stop pin. Also, you should look for corrosion pitting on the rotor.

The stator will need to be inspected by looking through it. Some pump suppliers will utilize plug gauges to check for wear or swelling of the stator. I believe that by far the best and most accurate means of checking for wear is by performance testing the pump on a test stand.

The plug gauges will only tell you if the stator is worn on the minor diameters where a pump test gives overall results.

During a visual inspection, you will need to look at the surface of the stator elastomer noting if the elastomer is dull or shiny. Are there any cracks on the surface of the elastomer? Are there any blisters or bubbles? Is the elastomer hard or soft?

The description of all of the above and the appropriate action that should be taken will be discussed in the next chapter.

Drive Assembly Operation and Maintenance

For the most part, the drive assemblies that are in the marketplace, are simple, consisting of two bearings—a radial bearing and a thrust bearing. Some drives include a right angle shaft with a 2:1 or 5:1 gear reduction ratio. Also, most drives have a backspin retarder or brake.

The bearing housing is an oil bath type configuration with the oil lubricating both the bearings and the brake.

The drive assembly also includes a stuffing box arrangement that utilizes cut packing to seal around the polish rod to prevent fluids from leaking.

Before operating the equipment, you should check to see that the oil level is correct by looking at the oil sight glass. Also check to see that the packing has been greased. The maintenance of the drive assemblies consists of changing the oil, and greasing the packing and top bearing in some cases. You should consult the respective drive manufacturer for instructions on maintenance and on the type of oil and grease to use.

However, most drives that are oil bath require an oil change about every six months using 85W:140 oil or the equivalent. If a drive assembly with a right angle shaft is being used, it is recommended that the oil change occur every 90 days. This is due to the possibility of metal shavings gathering from the spiral bevel gears. In a right angle drive, it is recommended that a synthetic oil be used.

The oil levels should be maintained between oil changes.

The packing area or stuffing box will need to be tightened occasionally if leakage occurs. It is only necessary to tighten the stuffing box cap by hand.

You will need to grease the stuffing box upon startup, then once a week. It is a good practice to wipe away any old grease before pumping new grease into the packing.

It is recommended that the packing rings be changed every six months. Consult the manufacturer for particulars.

Some driveheads are manufactured with a bushing and gland follower in the stuffing box arrangement to help centralize the polish rod to assure minimum amount of fluid leakage. Also, there are a variety of different types of packing available for different types of well fluids and operating conditions.

As part of your regular maintenance, you should check the tension on the belts if belts and sheaves are being utilized. All nuts and bolts should be checked on the motor mounts, stands, or supports and also on the belt guards.

If all the manufacturer's recommended maintenance programs are followed, you should be able to experience long run times from the PC Pump and drive assembly.

CHAPTER 7

TROUBLE SHOOTING THE PROGRESSING CAVITY PUMP

Due to the simplicity of the Progressing Cavity Pump and its low maintenance requirements because of fewer moving parts, there is less to go wrong. Normally, if problems do occur, they are to application not equipment.

With the proper trouble shooting analysis performed, any problem can be identified and resolved. If an operator or his pump supplier is recording readings from the pump and well on a regular basis, many problems can be avoided. Providing accurate data in a timely manner is the best preventive action an operator can take to assure long run times.

TROUBLE SHOOTING PROBLEMS

One of the first considerations in trouble shooting the Progressing Cavity Pump, as with any bottom hole pump, should be how the conditions of operation have changed. As discussed earlier, a good pump supplier will have taken readings such as electrical power draw (amps) and fluid levels during the initial installation. First, you should compare the current data with that of the initial well information that was recorded. This will help eliminate some of the guesswork in trouble shooting the problem.

Before assuming that the problem you are experiencing is a down hole pump problem, be sure to check and eliminate all the surface equipment such as the drivehead (bearings, gears, and brake), prime mover, electric motor, power lines (low voltage), fuses, hydraulic drives (RPMs may be set incorrectly), and variable frequency drives (may be programmed incorrectly).

All of the surface equipment should be checked by your pump supplier before calling a service unit to pull the pump. This action could save the end user an unnecessary service rig charge.

PROBLEM A:

Experiencing no fluid flow at the surface with the correct polish rod speed.

1. *Sucker rod parted* • Check amp reading compared to previous readings. Or pick up rod string to determine rod weight.

2. *Hole in tubing* • Shoot fluid level, pressure up on tubing.

3. *Tubing has unscrewed or parted* • Try to tag the tag bar/stop pin with rotor before pulling tubing to confirm tubing being backed off.

4. *Rotor not engaged in stator* • Rotor may have been spaced out incorrectly or the stator may have been installed upside down with the tag bar/stop pin on top.

5. *Rotor run through stator* • Rotor installed too fast and hard, knocking out the tag bar/stop pin, and the rotor head is blocking the discharge end of the stator.

6. *The rotor is broken* • Check amp readings, pull and replace rotor.

7. *PC Pump is worn out* • The fit between the rotor and stator is excessive, allowing slip. Have pump performance tested. Replace if need be.

8. *Stator elastomer chunked out* • Chemical attack or overpressuring of stator or high temperature. Pull and replace stator with proper elastomer.

PROBLEM B:

Experiencing no fluid flow at the surface with polish rod speed slower than originally set.

1. *The rotor may be on the tab bar/stop pin* • Check amps and re-space rotor.
2. *The prime mover is undersized for the application or is damaged* • Check prime mover, change if necessary.
3. *The stator elastomer may be torn from rotor being installed too fast and hard.*

PROBLEM C:

Experiencing no fluid flow at the surface with the polish rod not turning and the prime mover speed is correct.

1. *Drive assembly experienced a failure* • Check bearings or gears. Replace if necessary.
2. *The rotor is running on the tag bar/stop pin* • Check spacing of rotor and re-space if necessary.
3. *Solids locked up the rotor* • Lift rotor out of the stator and flush. You may want to swab well before reinstalling rotor.
4. *The stator elastomer is swollen* • Check for chemicals or bottom hole temperature. Replace with proper elastomer.

PROBLEM D:

Experiencing fluid flow, but at a lower than expected rate, with the correct speed.

1. *Production rate data incorrect* • Shoot fluid level, monitor amps, check with original readings, slow pump down if necessary.
2. *Hole in tubing* • Check fluid level, amps, pressure, and tubing. Replace bad joint of tubing.
3. *Pump intake may be blocked* • Pull rotor and try to flush out tubing and stator; pull stator if necessary.
4. *Rotor may not be fully inserted into the stator, resulting in lower pressure capabilities* • Check rotor spacing; re-space if necessary.
5. *Stator elastomer swollen* • Check for chemicals, bottom hole temperature. Replace stator with other elastomer if necessary.
6. *Rotor and stator are worn out* • Pull pump and have it performance tested. Replace if necessary.

PROBLEM E:

Experiencing fluid flow, but at a lower than expected rate, with a slower than expected speed.

1. *Sheave/pulley size is incorrect* • Check calculations and resheave if necessary, monitoring fluid levels.
2. *Hydraulic drive or variable frequency drive is set incorrectly for a slower speed* • Reset and monitor fluid level.

3. *Elastomer swell* • Check for chemicals and temperature. Replace stator if necessary.

4. *Low line voltage* • Check voltage readings.

PROBLEM F:

Experiencing production at surface, but it is pulsating with the correct speed.

1. *The well has a high gas to fluid ratio (GOR)* • Make sure pump is set below perforations, increase submergence if necessary, and monitor fluid levels.

2. *Well is pumped off* • Check amperage pull and fluid levels. Slow pump down if necessary.

3. *Stator elastomer may be experiencing blistering or bubbles due to gas permeation or stator elastomer has started to experience a failure.*

PROBLEM G:

Experiencing pulsating production at surface with uneven or low speed.

1. *Rotor is running on the tag bar/stop pin* • Check amps and re-space rotor.

2. *Rotor/stator internal fit is incorrect* • Check pump performance data; have pump retested.

3. *Sand production causing pulsation* • Make sure you have correct pump model for application; you may want to sand pump well and move up hole.

4. *Stator elastomer may be swollen* • Check for chemicals, temperature. Make sure you have correct elastomer for the application.

PROBLEM H:

Experiencing excessive backspin or reverse rotation when the pump is shut off or locking down during operation.

1. *The back spin retarder/brake has failed* • Call pump supplier out and have retarder/brake replaced or repaired.
2. *The rotor is locking up in the stator due to solids or stator elastomer swell* • Check for solids and pull rotor out of stator and flush if necessary; change elastomer if swell is the problem after checking chemicals and temperature.
3. *Rotor/stator fit is incorrect* • Check performance test data and have pump retested if necessary.

PROBLEM I:

The polish rod is not rotating with prime mover running.

1. *The belts have been thrown off or broken* • Replace belts.
2. *Belts became too loose* • Tighten belts, replace if necessary.

3. *Sheave or pulley may be broken* • Replace.

4. *If drive assembly is right angle drive, the pinion shaft may be parted* • Replace drive unit or have repaired.

PROBLEM J:

Experiencing excessive packing leakage on drive assembly.

1. *Packing gland needs to be tightened down more* • Do not overtighten gland.

2. *Packing to be replaced* • Replace or add packing.

3. *Polish rod is worn excessively.*

4. *There is no packing in stuffing box* • Install packing.

PROBLEM K:

Experiencing oil leakage form under bearing housing.

1. *The lip seal under the bearing housing is worn* • Replace with new seal.

2. *Bolts on bearing housing may be loose* • Tighten bolts.

PROBLEM L:

Experiencing excessive noise on the surface.

1. *Belt guard may be loose* • Tighten all nuts and bolts.

2. *Belts may be loose* Check and tighten belts.

3. *Electric motor may be experiencing a failure* Run motor without belts being hooked up; replace or repair motor.

PROBLEM M:

Experiencing excessive noise from down hole.

1. *There may be a bent sucker rod* • Check rods next time pump is pulled.
2. *The alignment of the drive assembly or wellhead may be off* • Check alignment and realign if necessary.
3. *Excessive down hole torque causing rod rap-up* • Check rotor/stator fit or stator for swell and solids.

COMMON FAILURES

A PC Pump can experience many different types of failures. Some are more common than others. I will describe in detail the most common rotor and stator failures, with suggestions on how you may be able to avoid them. I will list the failures in the order of what I believe to be the most common to the less common types.

In most cases, a failure with a Progressing Cavity Pump that involves the rotor and stator, not the sucker rod string or tubing string, can result in having to replace the stator and sometimes the rotor.

Even under normal conditions in which the pump experiences gradual wear to the point that the internal compression fit becomes too loose to overcome the pressure required to lift the fluid, the stator will need to be replaced. Under normal conditions, the rotor will wear at a slower rate than the stator. The rotor can be rechromed if the base metal is not damaged, whereas the stator cannot be repaired, and becomes a throw-away item.

1. *Run Dry Operation*: This is one of the most common types of failures seen with the PC Pump. The pump experiences a run dry operation when it is starved for fluid or operated in a pumped off condition for an extended time period. The pump experiences internal friction heat build-up due to the lack of lubrication. This results in the elastomer surface burning or scorching to the point that the pump loses its seal lines starting at its intake (suction) end and progressing up to the discharge end. Upon losing the seal lines, the pump can no longer build enough pressure to lift fluid to the surface.

The appearance of the stator elastomer surface will look hard, cracked, and crumbly. The internal diameter (ID) of the stator will appear larger than normal. In some cases, the rotor may looked burned and have some rubber on it.

To avoid a dry run operation, it is important that the end user provide accurate production data via the application data form to the pump supplier.

In many cases, it is a good idea to set the pump's operating speed (RPMs) at a setting that will produce a lower volume of fluid than that which is desired. Then, after monitoring the fluid level, the pump supplier can increase the pump speed until the desired production rate is reached.

Another method of avoiding a run dry operation is to utilize some form of pump off controller. Depending on the type of controller, the pump will be protected from a run dry by the controller shutting the pump off when it's in a pumped off state or by slowing down the pump. Flow controllers will be covered in more detail in Chapter 8.

2. *Gas Permeation:* Gas permeation is another common type of failure that the PC Pump can experience. It occurs when gas saturates or invades the stator elastomer, resulting in blistering or bubbles. The blisters and bubbles are most commonly found directly under the elastomer's surface. The result of a gas permeation is that the pump's torque or internal friction will increase, causing sucker rod parts and, ultimately, the elastomer to chunk out, turning into a catastrophic failure. In other cases, the torque within the pump becomes so great that the rotor breaks, not the sucker rods. This is most likely due to the rotor being put under stress by the stator grabbing it and by the horsepower on the surface applying torque at the same time via the sucker rod string.

When looking through the stator, one will be able to see blisters or bubbles on the surface of the elastomer.

These bubbles can range in size from very small to large ones that block your view. Usually, in the case of large blisters or bubbles, the elastomer will break away, ripping the top layer.

In some cases, the blistering or bubbles do not become apparent until the stator has been sitting out of the well for a given time. To avoid a gas permeation, the producer will need to give the pump supplier good data on his well prior to the installation, so the proper stator elastomer can be used and the proper pump setting depth selected.

If gas is present in the well, the stator should be set below the perforations, allowing the gas to break out and move up the annulus, using the well bore as a gas separator.

If it is not possible to set the pump below the perforations, you should consider running a gas/mud anchor to screen out as much gas as possible. Gas/mud anchors will be discussed in Chapter 8.

3. *Swell:* This, too, is a common problem that can occur with a PC Pump. The stator can experience swell for a variety of reasons. The stator elastomer can swell due to the aromatics of the fluids being produced, treating chemicals being used in the well, and the bottom hole temperature.

The end result is usually the same, an increase in torque. As in the case of gas permeation, the internal friction will increase due to the swell. This again can cause problems such as sucker rod parting, rotor breaks, overloads or blown fuses in your electric box. In severe cases, if the tubing is not tightened to API specifications and the rotor locks up in the stator due to swell, the tubing can come unscrewed.

When checking for swell, you should look through the stator and notice if the (ID) internal diameter is smaller. The best way to check for swell before the pump is pulled is by monitoring the amperage. If the pump is pulled, you should have it performance tested, the results of which will give you the pump's torques.

To avoid stator swell, it is vital that the pump supplier be given all the well data before installing the pump. If aromatics are a concern, a fluid sample should be tested as well as any treating chemicals.

When selecting the pump size and materials of construction, the aromatics, chemicals and fluid, and bottom hole temperature should be considered.

4. *Hysteresis:* This type of failure is always a catastrophic failure of the elastomer.

A hysteresis failure occurs when the elastomer is overpressured, resulting in a retardation of the elastomer from too much internal friction. The elastomer has a mechanical breakdown and begins to tear away or chunk out.

This type of failure occurs when a pump is installed in an application that is beyond its depth capacity or if the rotor is not engaged fully, resulting in the portion that is engaged becoming overpressured.

Often, a run dry operation is accompanied by a hysteresis failure, since a run dry begins at the intake of the pump and progresses up to the discharge. Before the pump can burn up all the way, the top section becomes overpressured and experiences a hysteresis failure.

Looking at the stator, you will find chunks of rubber that have been torn from the tube and evidence of tearing of the elastomer.

To avoid a hysteresis failure, you must make sure that the correct pump is sized for the application and that all pressures are accounted for (flow lines, tanks, etc.)

5. *Bond Failure:* A bond failure is a true manufacturing defect and should always be covered under the manufacturer's warranty.

Bond failures occur when there is either a weak bonding agent (adhesive) or one that is not applied properly to the stator tube before the elastomer is molded.

The effect of a bond failure is that the stator's elastomer will come loose from the stator tube, resulting in high torques followed by a catastrophic failure (lock down of the rotor). In many cases, the elastomer will come out in large sections, sometimes intact, in contrast to a hysteresis failure, in which you experience chunking out of the elastomer.

When looking at the stator tube, you will see the inside of the tube will be clean, with no adhesive on the tube. It may be only in a small area or over a large area.

Avoiding a bond failure is difficult. Your best defense is to ask your pump supplier what percent of their stator production results in bond failures. For most manufacturers, the number is very low; however, it does happen on occasion.

ROTOR FAILURES

1. *Abrasion/Solids:* When the PC Pump is installed in applications where solids or abrasive fluids are present, you can experience wear on the rotor, resulting in a pump failure.

This can occur both in applications that produce a large amount or a small amount of solids, but with the pump operating at high speeds (RPMs).

When the rotor experiences wear from abrasion, the pump will lose its compression fit and begin to experience slip, making it difficult to overcome pressure.

In looking at the rotor when the pump has been operating in an abrasive condition, you will see wear that follows the rotor profile over the length of the rotor.

The key to avoiding this type of failure is the operator providing the pump supplier with well data prior to sizing a pump for the application. The pump supplier can then size a pump to handle solids by oversizing the pump for the application and operating a low RPM.

If the wear on the rotor has not damaged the base metal, the rotor may be rechromed.

2. Rotor Wear: In pulling a rotor out of a well, you will frequently notice that on the top 6 to 12 in., the chrome will be missing on the rotor profile. This is usually due to the rotor rubbing against the tubing immediately above the stator. Since this portion of the rotor is not engaged into the stator, it does not cause an operational problem; however, if you do not have the correct size tubing and the rotor is rubbing, there may be an increase in torque as well as vibration.

You should consult your pump supplier for specifications on tubing and rod size with each PC Pump model.

3. *Pitting/Corrosion:* If the pump is installed in an application where there is corrosion, the rotor can experience pitting. This process occurs when holes are formed in the base metal due to the corrosion, which will result in wearing the stator elastomer.

Due to the porosity of the chrome plating used to cover the base metal, the corrosion is able to attack it.

In looking at a rotor, you will see pitting that can range in size from small holes to large holes, depending on the amount and type of corrosion as well as on the length of operation.

In most cases, the rotors cannot be rechromed due to the pitting. However, I have seen operators fill the holes with epoxy, reinstall the rotor, and then continue their run times.

4. *Chrome Flaking:* Chrome flaking, like pitting, is a result of corrosion; however, unlike pitting, it flakes off. Due to the porous nature of the chrome, the corrosion is able to get underneath it and lift (flake) it off.

In looking at the rotor, you will see areas where chrome is missing, but the base metal is still good. In some cases, the flaking may be a manufacturing defect.

In both cases of pitting and flaking due to corrosion, you will need to provide the pump supplier with accurate well data. This information will help determine if stainless steel should be used, opposed to alloyed steel with chrome plating.

5. *Rotor Breaking:* As discussed earlier, the rotor of a PC Pump has been known to break. In almost all cases, this is due to high torques; however, in some cases, manufacturing defects may play a part.

We discussed the different types of causes for high torque—swell due to chemicals, temperature, blistering and bubbles, and the presence of solids. Another area where high torque can be present is within the pump design itself. I have seen designs in which the internal fit causes the pump to experience excessive torques. This becomes very apparent when the pump is operating at low speeds (RPMs), resulting in excessive vibration and parted sucker rods or rotors.

A quality pump supplier will provide you with new pump performance test results so that you will be able to see what the pump's torque is. Again, you ideally want a pump that has the lowest torques possible while still maintaining high volumetric efficiencies.

If you are using a good pump supplier, in some cases, he may be able to salvage a broken rotor by welding it. However, you would need to consult the supplier or manufacturer about this option.

CHAPTER 8

PRIME MOVERS AND OTHER ACCESSORIES FOR THE PROGRESSING CAVITY PUMP

Due to the variety of both drive assembly configurations that the manufacturers offer, almost any type of prime mover can be accommodated to meet the needs of the end user.

The flexibility of both the Progressing Cavity Pump and its drive assembly make it adaptable to just about any well condition and power source. The PC pump can be powered by utilizing electric prime movers, variable frequency drives, hydraulic drive units (both electric and natural gas), and traditional natural gas engines.

With the PC pump's wide range of application uses in the industry, it is run in conjunction with many different types of accessory equipment. Many types of traditional oilfield equipment that are used with conventional pumping methods can be used with the PC Pump, while some accessory equipment was designed especially for use with the PC pump.

The use of such accessory equipment is dependent upon each application; there is no standard system that fits all well conditions or applications.

Examples of some accessory equipment are sucker rod guides, anchors-catchers, gas separators, and flow controls.

In this chapter, I will discuss how these prime movers and accessory equipment are applied to the use of the Progressing Cavity Pump.

PRIME MOVERS

Electric Drives

In my opinion, the first consideration should always be to choose electric power because overall it is the most efficient and cost effective means of producing an oil or gas well when maintenance is taken into account.

Electric prime movers make for a very clean, efficient, and smooth running installation. Electric motors of all types and sizes can be utilized with all of the drive assemblies offered by manufacturers (direct drives, right angle drives, etc.).

Many end users elect to utilize high-efficiency motors to receive some added power savings.

The electric motor is mounted in either a vertical or horizontal position, depending on the type of drive assembly being used. Sheaves and belts are then used on both the electric motor shaft and drive assembly shaft. The RPMs are achieved by changing the sheave ratio to meet the well's production rate as well as utilizing different RPM electric motors.

If an application calls for a slow-speed operation, a 1,200-RPM motor is used, whereas on a high-speed operation, an 1,800-RPM motor would be utilized.

One other big advantage to using electric motors is their noise level. They operate at a very low noise level with the Progressing Cavity Pump.

The electric motor with the PC drive assembly results in a very compact, low-profile unit on the surface when compared to other lift methods.

I am aware of one drive assembly in which the electric motor is coupled directly to the gear box. This eliminates the sheaves and belts as well as the motor mount and belt guard. Also, this drive makes for a low-profile unit.

Variable Frequency Drives

The variable speed drives have become popular with some users of the Progressing Cavity Pump. They are being utilized in two applications: when the well's flow rate is susceptible to change and when the production practice is to keep the well in a pumped off state and automation is utilized with the variable frequency drive.

The purpose of the variable frequency drive with a PC Pump, as with any other pump, is to have the flexibility to adjust the pump's speed without changing the sheaves and belts.

Since the cost of variable frequency drives has dramatically decreased in price recently, it is economical to utilize them on small horsepower applications. Even though the cost increases as the horsepower increases, it still offers great benefits.

Another big advantage of electric prime movers over the non-electric type is the maintenance factor. Other than checking the belt tension, there is very little maintenance required on an electric prime mover, whereas with hydraulic power units or natural gas prime movers, there is a lot more maintenance.

Hydraulic Drive Units

The hydraulic power units offer a flexible means of achieving variable speed in areas where electric power is not readily available.

The manufacturers of the Progressing Cavity Pump offer hydraulic drive assemblies to accommodate hydraulic power units.

The hydraulic units are offered with three-, four-, and six-cylinder natural gas engines, which run on the well's natural gas; this, in turn, powers the hydraulic motor that turns the PC drive assembly.

These power units are typically utilized in remote areas, but are also available with electric motors.

Besides the variable speed, another benefit of the hydraulic power unit is its ability to monitor torque through the hydraulic pressure. Some hydraulic power units are offered with automatic shutdown systems, which stop the Progressing Cavity Pump when it is in a pumped off state. The system utilizes its torque sensing ability to shut down the pump when its torque (internal friction) changes due to lack of lubrication and will restart the pump after a predetermined time period.

Automated oilers are available as well for the hydraulic power unit to help reduce maintenance.

In some areas such as northern Michigan, the hydraulic power units are installed in an environmentally correct building enclosure. The buildings not only keep the hydraulic units out of sight, but the producers also utilize them for their gas meter runs. The building also helps reduce the noise level and some contain a secondary spill container liner. The hydraulic power unit has a spill container of its own.

The hydraulic power unit requires more maintenance than the electric motor. Maintenance is required on the natural gas engine (checking oil levels, filters, etc.) and on the hydraulic unit (the hydraulic fluid level must be checked).

However, the hydraulic power unit offers a variety of benefits in one unit that you do not get from other prime movers.

Natural Gas Engines

In remote areas where electric power is not available, some producers utilize standard oilfield natural gas engines to power the Progressing Cavity Pump.

They use a drive assembly with a right angle shaft, belts and sheaves, and the natural gas engine.

However, this system does not offer any of the benefits of the hydraulic power unit, such as variable speed, torque monitoring, and shutdown systems.

Like the electric motor, to adjust the pump's RPMs (speed), you have to change the sheaves and belts. Alignment is very critical with a natural gas engine and a right angle shaft drive assembly. If the alignment is off or if too much tension is pulled on the belts, the right angle shaft can experience overhung loads, resulting in a shaft break.

Another area of concern with a natural gas engine is the vibration and the center distance from shaft to shaft. If the engine is set too close to the drive assembly, excessive vibration can result.

In some areas, I have seen the natural gas engine directly coupled to the drive assembly via a jackshaft. This design is used to avoid the overhung loading problem.

Again, with the natural gas engine, there is a higher level of maintenance required.

Generators

In areas where the producer does not currently have electric power, but expects to have it sometime in the future, he uses a generator to power the electric motor on the Progressing Cavity Pump.

The generator has a natural gas engine that uses the well's natural gas or other fuels. The advantage to this method is that the producer will not have to change his drive assembly, mounting, and belt guard from a natural gas unit to electric after electric power is available. He will have everything in place.

A few pump suppliers will lease generators for this purpose.

Sucker Rod Guides

In some applications, it is necessary to apply guides to the sucker rod string. The sucker rod guides are used to assist in reducing sucker rod and tubing string wear.

A great variety of rod guides are available from several manufacturers, some of which are now manufacturing guides designed for Progressing Cavity Pumps.

Documented reports from field experience indicate that rod guides have improved the life of the rods via fewer rod parts.

The types of applications in which rod guides should be utilized are high-speed applications, applications with dog legs or deviation, and applications that are experiencing sucker rod parting due to excessive torque created from some pump designs.

The most common type of guide used in PC Pump applications is the snap-on guide. Typically, there are no more than three guides per rod: one by the pin end, one in the middle of the rod, and one at the box end. The smallest number is one rod guide per sucker rod close to the coupling.

This type of guide is very good in high-speed applications.

Another type of guide that has been used with PC

Pumps is the molded on rod guide, which is molded permanently to the sucker rod. This type of guide has not been proven to be as successful as the snap-on guide and is more expensive. The molded on rod guide tends to create more drag and does not perform as well at high speeds.

Some manufacturers have developed what is referred to as a spin through molded guide for PC Pump applications. These are two-piece guides, one piece is molded onto the rod and the other piece is a sleeve that either snaps onto, or is welded onto, the molded piece. The idea is that the sucker rod and the molded piece will rotate while the sleeve will stay stationary.

There have been mixed results with these types of guides. I have seen cases in which the sleeve piece wore out or the weld did not hold together. I also have seen cases in which the molded piece wore out at a rapid pace. I would not recommend operating this type of guide at high speeds.

Another type of rod guide that is on the market is a coated rod coupling. This guide utilizes a machined sucker rod coupling with a molded coating. The coupling length is the same as a standard sucker rod coupling. This type of guide is good at keeping the rod coupling centralized; however, it should be used in conjunction with other sucker rod guides such as the snap-on so that the body of the rod is protected in application with deviation. Coated rod couplings have also been utilized in applications by themselves with success. On the other hand, there have been reports of some failures due to excessive wear and/or the coatings have come loose.

You will want to check the references of the manufacture to confirm their run times.

Another type of guide that has proven to be very beneficial and successful with the Progressing Cavity Pump, is the rod centralizer. This guide has the same design principle as the molded on spin through design that was discussed earlier, but with greater success since the rods rotate while the sleeve remains stationary.

The big difference and additional benefit to this rod centralizer over that of the coated coupling is its spin through design and its length.

The rod centralizer is a two-piece guide with a machined centralizer shaft that is typically 8.5 in. in length and a sleeve that is a molded nitrile material with a straight vane design. The sleeve fits onto the centralizer shaft and has an overall length of 4.8 in.

The design principle is that the sleeve becomes non-rotational, while the rod string is centralized and rotating. The sleeve vane design allows for a minimal amount of friction line losses in the tubing string.

This rod centralizer guide is often used in conjunction with a snap-on guide to assist in keeping the entire rod from rotating on the tubing string.

Depending on the application, the centralizer can be used between every rod in the sucker rod string or only at critical contact points.

I have seen rod centralizers used in applications where excessive sucker rod whipping occurs due to high torques in the PC Pump type and model being used. In one application, parting was taking place every two to three weeks where the centralizers were installed on the top ten sucker rods down from the surface and also on the bottom ten sucker rods, starting one rod up from the pump's rotor and the rod parting problem ceased. The end result was less down time and lower lifting cost, which prevented the end user from pulling the PC Pump out and going back to a beam pumping unit method. The end user plans on replacing the PC Pump when it wears out with one that is "matched" to his well conditions.

The key to successfully installing and operating any of the above guides with a Progressing Cavity Pump is to seek the expertise of your pump supplier and/or that of the guide manufacturer.

In certain applications guides should be utilized (such as deviated well, wells with dog legs, and high-speed applications); however, in other applications guides are not necessary.

Also, you will find that some guides are better suited for a particular application than are others. This is the reason you should ask the pump supplier for a customer user's list to check references. Be sure to ask about run times, type of problems encountered, if any, and service received.

ANCHOR-CATCHERS

On some occasions, an application may call for a tubing anchor to be used in conjunction with the Progressing Cavity Pump stator and tubing string.

A number of tubing anchor manufacturers have conventional anchors that are designed to prevent all vertical movement of the tubing string at the anchor during a pump cycle.

A few manufacturers have designed anchor/catchers for use with the Progressing Cavity Pump. These anchor/catchers are designed to not only prevent all vertical movement of the tubing at the anchor, but also prevent parted tubing from falling down hole. Although this is not a common problem, the tubing can back off and drop down hole.

If the tubing threads are inspected and are in good condition, and the tubing string is tightened to API optimum specification, there should not be any trouble with unscrewing. Generally, the prime mover being used will not have enough torque capability to unscrew the tubing.

Applications in which you may need to consider utilizing an anchor/catcher are those in which you anticipate high torque due to a heavy sand production, elastomer swell, heavy oil applications, or any place that unscrewing tubing is a concern.

Consult your Progressing Cavity Pump supplier for recommendations your application calls for a tubing anchor.

GAS SEPARATORS

As discussed in earlier chapters, the Progressing Cavity Pump will not gas lock like a conventional bottom hole pump due to the fact that the pump has no valves to stick. The pump efficiency can be affected by gas interference.

Also as discussed earlier, when the PC pump is being applied in an application where gas interference is a concern, the pump should be set below the well's perforations or pay zone where possible. This is referred to as natural gas separation. By setting the pump below the perforation, the gas is allowed to separate and break up the well's annulus as opposed to passing through the pump's intake.

In applications where there is no rat hole below the perforation or where there are other concerns making it impossible to set the pump below the perforation, a gas anchor can be utilized to assist in gas separation.

The most common gas separator or anchor is referred to as a "poor boy" gas anchor. It has been utilized by producers for several years on conventional bottom hole pumps to deal with the problem of gas interference and separation.

The poor boy gas anchor design consists of a tubing pup joint with perforations or slots at the top of the joint. A dip tube usually 1 1/4 in. in diameter runs inside the pup joint and has perforations at the bottom of the tube. The pup joint is attached to the bottom (intake) of the pump.

The fluid enters through the opening at the top of the pup joint and reverses its flow down to the opening in the dip tube, at which time the gas separates from the fluid.

The gas is able to break out and move up the well's annulus, and the fluid then enters the pump's intake through the dip tube.

Another type of gas anchor design uses the poor boy design but places cups around the perforations at the entrance of the pup joint to create a change in the fluid velocity, allowing for more separation and making the gas anchor more efficient. However, if the application is expected to produce any solids, you would not want to use the cup type gas anchor.

FLOW CONTROLS/ PUMP OFF CONTROLS

In the SPE paper by J. D. Clegg, S. M. Bucaram, and N. W. Hein, they discussed the fact that the Progressing Cavity Pump performs poorly on a time cycle and on pump off controller application. In addition, they stated that pump analysis is based only on production and fluid levels, and that dynamometers and pump off controllers are not possible to use.

I will say that collecting production and fluid level data play a big role in performing pump analysis on a PC pump, as they do on other artificial lift methods.

However, since the time the SPE paper was written, the use of pump off controllers with the PC pump has been perfected. In fact, the technology has advanced to a level that provides a fully automated pump controller, which I will discuss in this section.

Since the Progressing Cavity Pump stator's materials of construction include an elastomeric component, there is the possibility of experiencing a "run dry" operation.

There are applications in which it is recommended that a flow control device be utilized to protect the PC Pump from burning up. For example, in an application in which low production rates are expected (100 BFPD or less) or the production of the well can vary without notice, some water floods could experience this result when injectors go down. The other application in which a flow control device could be considered is one in which the fluid level in the well is going to be maintained at or near the intake of the pump.

It is very important to keep in mind that a Progressing Cavity Pump can experience a run dry operation or burn up in a very short time period if there is no fluid entering the pump's intake to lubricate the pump, lowering the friction that causes heat build-up. When the pump is starved for fluid, the friction between the rotor and stator increases, resulting in internal heat build-up. Thus, it is critical that the pump is not operated in a pumped off state for any extended time.

There are a variety of devices on the market that

can be used with the Progressing Cavity Pump. They vary in quality, effectiveness, and price, and also range from the very simple to high tech. The following is a brief description of the devices. You will want to consult your pump supplier to see what types of controls he offers and to discuss if they are necessary for your application.

One of the simplest and most inexpensive methods to monitor flow is with a pressure switch. This flow control is mounted in the well's flow line and it monitors the high-low pressure of the liquid in the flow line. You are able to have a preset high-low pressure setting so that when the pressure hits that point, the switch can shut down the pump. The problem, however, with this type of gauge is that it has to endure any vibration that occurs at the wellhead.

The biggest drawback to this type of control is that its reaction time is too late. The pump can be in a pumped off state down hole while the flow line pressure may not have dropped yet. So, its accuracy is open to scrutiny.

Along the same line, there is a control that uses differential pressure readings in a switch to monitor flow, also opposed to reading the flow line pressure only.

This switch utilizes an orifice plate to create a pressure drop or differential. The preset point is programmed at a pressure that is less than that of the normal operating pressure. If a drop in the differential occurs that is lower than that of the preset point, then the pump is shut down.

This device is less likely to have problems due to vibration even though it is also mounted in the flow line. Best of all, this device will monitor the flow line pressure with more accuracy.

Another control that has been applied to the PC Pump as a pump off device is a flow meter. The principle behind this design is to monitor the fluid flow rate at the surface, sensing any change in it.

It allows you to calibrate a low fluid flow set point and if the production rate drops to that point, the meter is able to sense the change and shut off the pump.

Again, this type of control is mounted in the production flow line so that vibration can occur; however, this type of device is able to endure the vibration. The cost of this unit is higher than that of the pressure controls.

A thermal dispersion device uses two probes that are installed in the flow line, allowing fluid to pass through and over the probes. The thermal probes monitor the temperature variance between them. The temperature is then correlated to a measured milliamp signal and when a low one is read, it indicates a change in the fluid production rate. You can have a preset, low milliamp set point that when reached can shut down the pump.

The problem with this device is its accuracy. There are a lot of variables in the fluid make-up alone that can affect its accuracy. Another concern is that since the device has not been widely used with the PC Pump to date, its reliability in this application is not proven yet.

As discussed in an earlier chapter, if a producer is utilizing a hydraulic drive unit, it can come equipped with an automatic shut down device.

By monitoring the down hole torque via its hydraulic pressure, the system can sense a change in the internal friction of the pump. The system has a set pre-established torque level; when the torque passes this set point, the system will shut down.

Another feature of this system is full restart capability. This means that the system can overcome the starting torque of the pump, which can be higher than the pre-established setting point.

This system also offers a programmable time delay so that you can set the amount of time you would like the system to stay shut down before restarting.

The newest and in my opinion, most accurate controller on the market is an Automated Pump Controller that is used on electric applications. Like the hydraulic automated shutdown system, the automated pump controller (APC) monitors actual bottom hole conditions that affect the performance of the pump and makes the appropriate adjustments.

The APC monitors well conditions and uses variable speed technology to operate the PC Pump at the optimum speed, based on preprogrammed instructions and current well status.

The APC monitors the change in electrical current that results from an increase in internal friction when the well condition changes.

If the well condition changes, the APC slows down if the well's fluid inflow decreases. On the other hand, if the fluid inflow increases, the APC increases the pump speed. And, most importantly, if the pump starts to operate in a pumped off environment, the APC will shut down the PC Pump and cycle back on after a predetermined period. Due to the new technology that has been developed, the end user is offered a PC pumping system that is totally automated and protected. Not only is the producer able to optimize the pump's volumetric efficiency, increase production, optimize the pump's mechanical efficiency, and receive longer pump life, but he is also able to lower operating costs.

In an application that requires a pump controller, by utilizing the Automated Pump Controller, the producer can protect both his PC Pump from a run dry operation and himself from the costly replacement of a new PC Pump stator and the cost of the service rig time.

There is no other flow control and/or pump off control on the market that monitors down hole conditions and is capable of changing the PC Pump's RPMs to match the actual current well conditions. Thus, the APC should be a consideration for any producer who has concerns about dry run operations.

If you have concerns about protecting your pump from a run dry operation, consult your PC Pump supplier and discuss your application to see what type of device he has to offer or seek out the supplier who offers the best expertise and technology.

Anti-Friction Unit

A new development in accessory equipment for the Progressing Cavity Pump to assist in problem application where friction is inherent is the Anti-Friction Unit or Lift Head. It is recommended in applications in which increased friction is imminent due to solids that have a tendency to fall back onto the pump during operation or when the pump is shut down. It is also useful in applications in which elastomer swell takes place or in which the pump brand has inherent internal friction problems.

The Anti-Friction Unit consists of two hydraulic cylinders that are mounted on the PC Pump's drive assembly and are powered either by a hydraulic power unit that is operating the PC Pump or by a hydraulic pump that is powered by an electric motor, used for electric PC Pump applications. When the system senses an increase in torque (friction), the Anti-Friction Unit is activated, automatically alleviating the friction by physically raising the sucker rod string and the PC Pump rotor vertically. Then the device moves back down into its original position. The unit lifts the sucker rod string and PC Pump rotor while the rods and rotor are still rotating, allowing any solids that are present to be pumped and moved up hole through the tubing string.

This new technology is an excellent device for the remote area of some coal-seam well activity where there is an inherent problem with coalfines and it is very costly to have a service rig move in on a well to lift the sucker rod string and PC Pump rotor to unstick the pump.

The device could also be utilized in heavy tar sand applications, where it is common for the PC

Pumps to lock down due to sanding up.

This is yet another example of the acceptance of the Progressing Cavity Pump by the oil and gas industry, with qualified experts developing accessory equipment to accommodate the PC Pump in solving production problems.

CHAPTER 9

THE IMPORTANCE OF PUMP PERFORMANCE TESTING

Throughout this book the importance of pump performance testing has been stressed. I have found that many oil and gas producers have no idea how their pumps are truly performing. This includes progressing cavity pumps, conventional bottom hole pumps, and electric submersible pumps alike.

To assist the producer in optimizing his production, it is vital to have a pump performance test before the pump is installed into the well. With the new pump test data, a pump supplier or producer is able to compare the mechanical and volumetric efficiency of a pump that has been in operation for a given time period to the new test performance.

Since the key to a successful PC pump installation is "matching" the producer's well conditions, it is necessary to have a new pump test performance to find the most efficient rotor and stator combination.

After a pump has been in operation for awhile, if the operator pulls the pump for any reason, such as low volumetric efficiencies, sucker rod parts, or to treat the well, he should have it retested. Retesting is necessary to check for wear and loss of efficiency. This precaution can result in saving the end user the cost of installing a pump that is worn out or close to being worn out.

INITIAL PERFORMANCE TESTING

Another point that has been stressed throughout this book is that an end user needs to select a pump supplier who not only offers pump performance testing but also matches rotors and stators to a given set of well conditions, and supplies those test results to the end user.

There are a few reasons for stressing this. For one, it has been my experience that PC pump manufacturers do not fully understand the importance of pump performance as a whole when it comes to applying the pump to a down hole condition.

I found that a few manufacturers claimed to do 100% testing of their pumps prior to shipment, but in fact, they would test only the stators of a particular model with one "test stand rotor" of a similar model.

Then they would check the crest-to-crest diameters of the rotors as they were manufactured to see if they were within specifications for that model. The result is that the end user could purchase a stator that performed to a certain specification with the test stand rotor, but when installed into a well with a different manufactured rotor its performance would be totally different. There may be two reasons for manufacturers taking this approach: the cost of time and manpower involved in testing rotors and stators in matched sets, and understanding the applications. Many manufacturers and some pump suppliers believe that if the product is being manufactured within specifications, only a small variance in performance will occur, particularly after thermal expansion of the elastomer.

I find fault with this belief because I have seen from measuring the crest-to-crest diameter of some manufacturer's rotors that not only can there be a big difference between diameters from one rotor to another in a particular rotor model, but also the length of some manufacturers' rotors vary a great deal, many times outside their own specifications.

Secondly, I do not agree that after installing the

pump in an application one will not notice the variance in performance. It is so critical to have a PC pump with the best mechanical efficiency possible while maintaining very good volumetric efficiency that if you are relying on thermal expansion to tighten up your pump to achieve efficiencies, you most likely will have a PC pump that will also experience high torque problems and vibration. Also, if the thermal expansion does not occur, you will end up with a PC pump that has poor volumetric efficiencies. Again, this misunderstanding in performance is related to the particular manufacturer and/or supplier not fully understanding the pump and its application.

Another point of concern is that some manufacturers and their suppliers who do performance testing of the rotors and stators do not always understand what to look for in a totally efficient pump.

It is important to take into account all of the well conditions, and to performance test the rotor and stator at the fluid temperatures and operating pressures under which the pump will be operating.

Looking at the pump performance test result in Figure 9-1, you can see that the pump supplier checks the pump's performance in three areas: (1) mechanical and volumetric efficiencies, (2) torque, both measured and theoretical, and (3) production by pressure

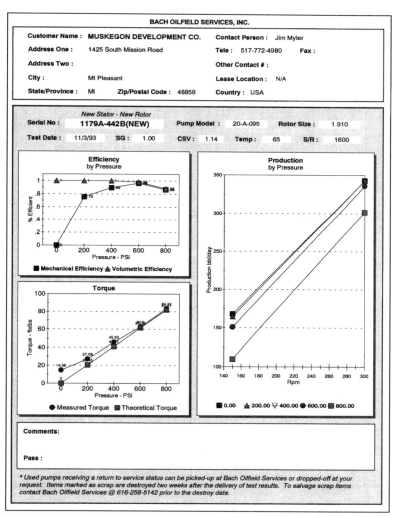

Fig. 9-1

at a given RPM, all in a curve format.

It is vital to achieve maximum mechanical efficiencies at the maximum operating pressures at the operating temperature. As you can see in Figure 9-1, the pump's efficiency changes at different operating pressures. Also, keep in mind that since one of the Progressing Cavity Pump's stator materials of construction is of elastomer, the pump element's consistency in efficiencies will vary from stator to stator within a set range of specifications due to the manufacturing process, usually the curing process of the elastomer. This lends even greater support to the practice of matching rotors and stators to a given set of well conditions. This also confirms the realization that the manufacturer is a parts supplier, i.e., rotors and stators need to be performance tested in a matched set to create a pump.

Another important feature to look at in the new pump performance test is the internal friction of the pump. By taking the difference between the measured torque and the theoretical torque, you will obtain the PC pump's actual internal friction. This has a direct effect on mechanical efficiency and horsepower required to operate the pump. Other benefits of low internal friction in the pump are longer run times from fewer problems such as sucker rod parting and excessive down hole vibration, both of which can occur due to high torque.

The main goal of the initial performance test is to achieve long run times that will reduce the producer's lifting cost. This goal is accomplished by providing the producer with a Progressing Cavity Pump that is not only of superior quality but also gives the best possible mechanical efficiencies while maintaining very good volumetric efficiencies. This will result in a pump that will operate trouble-free, keeping down operating costs.

The Progressing Cavity Pump would have experienced much greater success over the last 12 years or so had the manufacturers and/or pump suppliers taken this approach of using initial pump testing and more accurately applying the product.

Again, make sure your pump supplier knows what to look for in a pump test and provides you with the pump test results.

PUMP INSPECTION AND RETESTING

Pump inspection was briefly covered in Chapter 7 in the section on trouble shooting.

After it has been determined that the rotor and stator need to be pulled out of the well, the pump should be inspected to check its condition. In inspecting the pump, the supplier will check for wear on the rotor and stator both by visual inspection and by measuring the pump's diameters.

In looking at the rotor, one can see any signs of

wear on the rotor surface such as from abrasion. Abrasive wear can follow the contour of the rotor, whereas normal wear is more evenly distributed over the entire rotor. At the time of inspection, the supplier will need to measure the rotor's crest-to-crest diameter to determine if wear is present.

Also, the supplier will look at the rotor to see if there has been any corrosion attack on its surface, which could result in enhanced wear. As discussed earlier, due to the porous nature of the rotor's surface (chrome), it is possible for corrosion pitting to occur and attack the base metal. At this time, the supplier will need to determine the severity of the damage and whether or not the rotor needs to be replaced.

The pump supplier will also look at the bottom of the rotor to insure that it has not been run on the tag bar/stop pin as well as take note of any marks on the rotor that look like discolorations, burring, or burnt rubber on the rotor indicating a run dry operation.

In some cases, the supplier will note that the rotor has wear on the top portion where it has been running outside of the stator and has been running on the tubing, resulting in the flattening of the rotor crests. This rotor can be rerun as long as the portion that is worn is not installed in the stator.

As discussed earlier, the on site stator visual

inspection is limited to looking for signs of a run dry by seeing if the stator internal diameter appears to be larger than normal as well as determining if the elastomer is hard and crumbly or cracked.

The effects of overpressuring will result in chunks of rubber or large, loose pieces of rubber in the stator.

Another visual sign you can look for in the stator are the effects of swelling, checking to see if the stator's internal diameter appears to be smaller than normal. You can also take note of the texture or durometer of the stator's elastomer.

While looking through the stator, you will be able to see blistering or bubbles if the gas permeation is present. The bubbles range in size from very small to large enough to block the opening. This result may depend on how long the stator has been out of the well and how much permeation has taken place.

The pump supplier may want to take note of any unusual smells in the stator. After a run dry operation, one is able to smell burnt rubber or if there have been chemicals present during pump operation, you may be able to smell their effects.

Retesting

Retesting the PC pump is a necessity. Any time the pump has been pulled after a long run time or a problem has been reported, it should be retested to compare its current performance with its original performance.

One argument for not testing and selling the PC pump in a "match" set is that if one of the pump's elements (the rotor or stator) experiences a failure and you replace it, you either have to replace both elements or you end up with a pump that is not matched.

The proper procedure is to inspect and test the used element with the new replacement part because after a period of operating time, the rotors and stators can experience operating wear that changes the contour of the pumping elements. First of all, you want to make sure the surviving pump element can be reused and then have it retested with the new element to see if the pump performance is satisfactory for the well conditions.

The result may be that the combination of the used part and the new part is unsatisfactory and should not be run into the well. In this case, the time and money spent on a pump performance test paid for itself by saving the operator the expense of installing a pump that would not perform to the desired efficiencies. In the case that the elements performed satisfactorily, the operator's money was well spent to insure a successful pump installation.

Most pump performance tests will cost the producer around $200.00. This is a small amount of money to spend compared to the cost of a service rig to install the pump so you want to make sure you are installing a pump that will operate satisfactorily. When it comes to performance testing a pump that has been pulled out of a well for other reasons than a major failure, the money spent on the performance test is good insurance to check efficiencies in order to ascertain whether or not the pump can be reinstalled.

An example of a before-and-after pump performance test can be found on the following page. In Figure 9-2, you will find the results of the initial performance test; in Figure 9-3, you will find the results of the pump performance test ten months later.

The results of this particular pump manufacturer's type and model indicated that very little wear occurred. In looking at the volumetric efficiencies, you will see that even after a 10-month run it is still at 98% efficiency at 600 psi of pump lift. During this time, the measured torque of the pump has also changed very little, (65 ft. x lbs. on the initial test vs. 66.6 ft. x lbs. on the retest) and this amount of change could be a result of the accuracy variance of the test stand equipment itself.

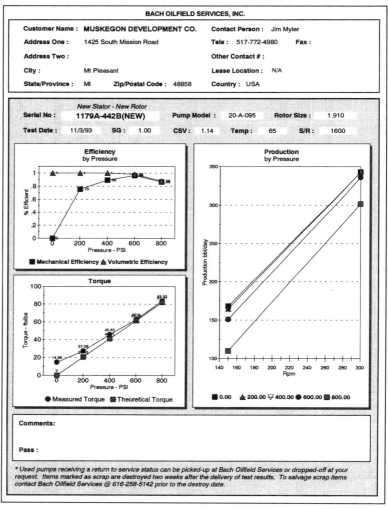

Fig. 9-2

The end result is that this operator can confidently spend the money to reinstall his pump with the assurance that he will receive the performance that he deserves and that the pump supplier promised.

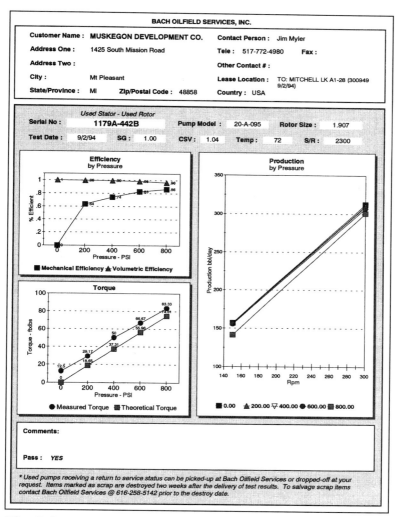

Fig. 9-3

As you can tell, the information that can be supplied to the end user as a result of the pump performance test has great benefit and every Progressing Cavity Pump operator should utilize this resource.

CHAPTER 10

MEETING THE FUTURE ARTIFICIAL LIFT NEEDS WITH THE PROGRESSING CAVITY PUMP

With its more than 60 years of solving application problems worldwide in a variety of industries, the Progressing Cavity Pump has continued to develop to meet the needs of many pumping applications.

The Progressing Cavity Pump has only been applied to artificial lift applications in the oil and gas industry for 15 years. However, during this short time period, the product line has continued to grow and develop at a rapid pace. When the Progressing

Cavity Pump was first introduced to the oil and gas artificial lift market, there were only two pumps being offered to the end users, with a production range of 100 BFPD from a depth of 2,000 ft. If you turn to Chapter 3, you can see just how much growth in the product range has occurred.

The manufacturers and their pump distributors have responded to the changing needs of the industry by developing products with wider capabilities.

A lot of the development of accessory equipment has come from the pump supplier or distributor due to their efforts to meet the needs of their customers in a variety of marketplaces. In fact, in many cases the pump supplier has been the forerunner in the advancement of the Progressing Cavity Pump by bringing to the manufacturers not only the needs of the industry, but also, in some cases, the solution.

As the industry's needs continue to change, the Progressing Cavity Pump still has room for growth and advancement, not only through continued research and development, but also through education of the oil and gas industry on how to successfully apply it to their applications.

INDUSTRY NEEDS

Anyone who has been in the oil and gas industry for any length of time has come to know that it is a very volatile industry. As the economic condition changes, so does the focus of the end user.

For the past five years or so, the industry has had few new well starts and marginal oil wells have been plugged. The greatest amount of new activity in the United States has been in the area of natural gas wells, including the coalseam activity. Until recently, the natural gas prices have been less volatile than oil prices. The activity in coalseams was assisted by the tax credit on unconventional gas, which has since expired.

All of these conditions lead to one hard fact—the oil and gas producers have to reduce their operating costs. The Progressing Cavity Pump is able to help the end user achieve the goal of lower lifting cost. If the end user selects the proper manufacturer and pump supplier, he will end up with a PC pump that will help lower his lifting cost by providing longer run times, resulting in lower maintenance on the well. The PC Pump is proven to be a more efficient method of artificial lift, resulting in lower electrical cost and high productivity.

If the proper supplier is selected, the end user will receive a PC Pump that provides him with the ability to equip his well for less capital cost while optimizing his production and experience lower operational expense with no unexpected or unnecessary workovers, resulting in increased net profit.

By far, the greatest needs of the industry are in the areas of education and expertise. In order for the continued growth of the Progressing Cavity Pump, the industry needs quality pump suppliers with the knowledge and expertise to apply the product to applications, resulting in a successful, long, problem-free run time.

A qualified pump supplier will be able to assist the end user in identifying the solution to his production needs. The end user needs to be educated on the proper selection and operation of a PC Pump. If the qualified pump supplier is doing his job, he will educate the end user on the philosophy of sizing the PC Pump in matched sets for a given set of well conditions, as well as operating it at lower RPMs wherever possible to increase life and lower maintenance cost.

Another area of need is in the product range. The Progressing Cavity Pump has experienced a large amount of growth in the product range during a short time frame and will continue to broaden its range in the future.

Currently, there is a need in the industry for PC Pumps that have a greater production capability at greater depths. In recent years, the Progressing Cavity Pump manufacturers have increased the production range of the product line. This has made it possible for the end users to replace some electric submersible pumps with the PC Pump, resulting in large power savings and lower repair costs.

If the manufacturers are able to extend the pump's capabilities to higher volumes (over 3,000 BFPD) at greater depths (4,000–6,000 ft.), the producer will receive an even greater savings due to capital cost, lower power consumption, higher efficiencies, and lower maintenance cost.

In order to accomplish this, the Progressing Cavity Pump manufacturers will need some assistance from the sucker rod manufacturers.

This brings up another area that needs to be addressed. The industry is in need of published data on the effect of sucker rods in rotation. At some point, one of the sucker rod manufacturers will need to do the necessary research because an increasing amount of their product is being applied in a rotational operation and to date, no data is available discussing the effect on the sucker rod.

With this type of research, the rod manufacturers may be able to develop a means of extending the depth of PC Pump operations with a sucker rod string.

This type of research and data gathering should be demanded by the end users of the sucker rod manufacturers and the Progressing Cavity Pump manufacturers.

Lastly, the industry has a need for an elastomer that can operate in a broader range of applications. As discussed earlier, I have found that through proper matching of the rotor and stator, the PC Pump can operate successfully in applications that had problems that were previously blamed on swelling of the elastomer, but in fact, the fit of the pump was the real problem.

However, there is still a need for elastomers that can operate in very high API gravity (38°–45° API oils). Also, there is a need for development of elastomers that can handle H_2S and high-temperature applications.

The Progressing Cavity Pump manufacturers need to work closely with the oil and gas industry, as well as with their rubber suppliers to continue the growth and application range of their elastomers.

FUTURE DEVELOPMENTS

Product development should be an ongoing process in order to meet the changing needs of the industry in which it is used and to expand the application range of the product.

With the Progressing Cavity Pump being such an old method of pumping, there have been years of research and development efforts. However, with every industry and new application such as the pump being utilized as a method of artificial lift, there is a need for new research and development. Basically, what has worked in the past does not always work today. Many manufacturers have discovered that the methods of manufacturing industrial PC pumps have to be altered sometimes for down hole applications. Also, materials of construction such as elastomers used in the industrial applications do not always suit the needs of the down hole applications. Past research and documented data provided by the industrial industry can prove to be very beneficial to the oil industry.

Future developments in the use of the Progressing Cavity Pump for artificial lift are very important in order to continue to expand this method of lift within the oil and gas industry. A great number of applications in the industry could benefit from such development.

One development that the industry will benefit from in the very near future is a Progressing Cavity insert pump. This new development will consist of a PC pump in which the rotor is engaged in the stator prior to installation in the well and then installed on the sucker rod string alone and seated in a standard seating nipple like a conventional bottom hole pump.

The big advantage to this type of PC pump vs. the traditional type is that the stator will not need to be run in on the tubing string, resulting in lower installation and maintenance cost. The possibility of unscrewing the tubing should be eliminated as well.

The shortcoming of a PC insert pump is that it will have to be a low-volume pump due to the diameter restriction. The current designs that I am aware of will have a production capability of 15 BFPD per 100 RPM. By taking the philosophy of slow speed operations to receive maximum life with limited wear, this pump will be commonly used in applications of 30 to 45 BFPD at depths approaching 4,000 ft. and possibly deeper (up to 6,000 ft.).

Depending on the manufacturing cost, the pump supplier should end up with a total pumping system that is very competitive with both the traditional beam pumping units and bottom hole pumps, which are currently being utilized in these low-volume applications.

This is an example of the manufacturers and pump suppliers developing products to meet the needs of the oil and gas industry. There are many stripper wells or marginal wells in which the operator needs a product that not only has a lower capital cost but also is more efficient than the traditional equipment used in the past. In addition, the end user has a pump that is more versatile and requires lower power consumption.

Another ongoing development that the end users will be able to benefit from in the near future is an electric submersible Progressing Cavity Pump. This concept was developed and marketed a number of years ago by Robbins and Myers, Inc. prior to their formation of a division solely dedicated to the oil and gas industry some 15 years ago.

At the time Robbins and Myers, Inc. developed the electric submersible Progressing Cavity Pump, it only had a depth range of 1,500 ft. with a production range of 33 barrels per day.

The idea is to run a Progressing Cavity Pump down hole with an electric submersible motor and electric cable, creating a more versatile and efficient electric submersible pump.

This will eliminate the sucker rod string and hopefully make it possible for the PC pump to be run to greater depths in the near future.

The only problem with past designs was that the down hole motors ran at a very high RPM, enhancing wear on the rotor and stator. Also, down hole gear reducers were very expensive and had a larger diameter than was needed in this application. However, there have been some new developments recently with the gear reducers, indicating that some manufacturers are readdressing this application potential.

Besides eliminating the sucker rod string, other benefits of the electric submersible PC pump could be an ESP that is not only more efficient, resulting in lower horsepower required, but also better at handling gas in a solution and solids.

Continued research and field testing are necessary before marketing this pump to the end user, in order to insure a product that is cost effective and can provide long-term life.

As discussed earlier, the other area in which future development is necessary and ongoing is in achieving higher volumes of production with the Progressing Cavity Pump. The manufacturers are attempting to develop a higher volume PC pump while being able to operate within a 5.5-inch casing. One means of achieving this is by utilizing the multi-lobe design that is used in down hole mud motors for drilling operations.

As discussed in Chapter 1, the typical down hole Progressing Cavity Pump consists of a single external threaded helical gear rotor, which rotates eccentrically inside a double internal threaded helical gear stator of the same minor diameter and twice the pitch length.

This results in a 1:2 lobe ratio.

A multi-lobe design would have a 2:3 lobe ratio, resulting in a higher volume rate. The critical points of this type of design are internal friction and vibration.

The conventional multi-lobe elements for mud motor operations have high internal friction (torques), which is required for the drilling operation; however, for the use of production pumping, a low-torque design with very low vibration is needed. So, once again the internal fit of the pump is the key to a successful pump operation. This product development will take longer than that of the two previous developments discussed. However, it is a big priority of some manufacturers due to the need in the industry to find methods of producing higher volumes of fluids more efficiently.

There is still room for a lot of research and development of the Progressing Cavity Pump for the purpose of producing oil and gas wells. Again, this product is still very young in relationship to other methods of artificial lift equipment and has advanced at a rapid pace in the short time that it has been marketed as an alternative to the traditional methods.

Now that the manufacturers and their pump suppliers have a better understanding of the true needs of the oil and gas producer, the job of developing products to meet their needs has become easier.

So, with the continued product growth of the Progressing Cavity Pump, its application range will expand even further.

ISO 9001

With the increased importance now being placed on lowering lifting cost by the oil and gas operator, the quality of the products they use plays a greater economical role than ever before. The quality of the PC pump can assist the operator in achieving this goal by offering maximum efficiencies while maintaining durability and providing long life.

Some manufacturers are seeking out ISO 9001 as a means to show their commitment to their customers in providing quality control.

ISO 9001 is the internationally recognized and accepted standard and specification for Quality Management Systems. The end result of this standard is that the customer will receive conformance at all stages of manufacturing from the time the order is placed through all the phases of manufacturing, and up until the time of delivery.

ISO 9001 assures the customer that the manufacturing process will be consistent and there is documentation to support all phases of the operation.

By controlling the process and documenting the work being performed, a manufacturer is able to implement changes whenever necessary and insure that they will be carried out in a regulated way.

After a manufacturer has been approved for ISO 9001 certification, he has to maintain the control and documented system or he risks losing certification. Also, every three years the manufacturer has to undergo an audit of his quality management system.

One of the original licensees of the Moineau principle, Mono Pumps, LTD, became the first Progressing Cavity Pump manufacturer to achieve accredited approval to this standard.

This type of quality standard from the manufacturer along with a quality pump supplier who demonstrates the knowledge and true understanding of not only the PC pump principle but also of its application to a given set of well conditions [via supplying the rotor and stator in match sets, supplying performance testing, and operating the pump at moderate speeds (low RPMs)] will insure that the oil and gas producer will receive the efficiencies, low maintenance, and long life they need to accomplish the goal of lowering the cost of barrels lifted.

CHAPTER 11

CONCLUSIONS

The Progressing Cavity Pump has experienced steady growth during its long history as an efficient method of moving fluids of all types.

As an artificial lift method, it offers the oil and gas operator a cost effective, efficient alternative to conventional methods of lift.

When the operator has an application that requires lift equipment, he should consider using equipment that will provide the lowest operating cost possible. There are applications in which the operator could select lift equipment that can be purchased at low capital cost but its operating cost may be higher than that of other methods whose capital cost may be more expensive initially. Thus, an operator will want to take all his parameters and expectations into account when selecting a method of lift. He should look at the total operating cost of the equipment, not just the capital expense.

In the SPE paper by J. D. Clegg, S. M. Bucaram,

and N.W. Hein, they discuss the importance of not force-fitting artificial lift equipment to applications because this action could result in premature failures and problems.

On the other hand, they suggest that if an operator's application is on land, his first choice should be a beam pumping unit. I do not agree with this thinking. I see this as a force-fit. I believe that the operator should take each of his applications case by case, selecting the most efficient, cost effective method for his application.

In regard to the Progressing Cavity Pump, if an operator selects a pump supplier who takes a scientific approach to the application and equipment, such as pre-testing the pump, doing in-well pump performance tests, and retesting pumps, he will experience low operating costs and long life in his PC pumping system.

If you are an operator who has tried a Progressing Cavity Pump some time ago, you should take another look at them today. The pump suppliers as well as many operators have learned a great deal since the pump was first introduced to the artificial lift market, thus extending the PC Pump's application range and abilities to perform in a cost effective manner, saving the operator a great deal of money.

The greatest area of concern continues to be the education of the oil and gas industry in applying the Progressing Cavity Pump to artificial lift applications. A great deal of work still needs to be done in this area by all the PC pump suppliers.

REFERENCES

Bach Oilfield Services, Inc., Progressing Cavity Pump brochure and manuals.

Griffin/LeGrand, Screw Pump operating manual.

SPE 24834, J. D. Clegg, S. M. Bucaram, and N. W. Hein.

SPE 25448, K. J. Saveth.

INDEX

A
Acknowledgements ....p. x
Anchor/Catchers ....Ch. 8, p. 108, 109

B
Beginning as artificial lift method .....Ch.1, p. 1, 2, 3
Bucaram, S.M. ....Foreword p. viii,
                                Ch. 2, p. 12, 14
                                Ch. 8, p. 111
                                Ch. 11, p. 149, 150
                                Reference p. 152

C
Caisson pumping applications ....Ch. 4, p. 44, 45, 46
Centralizers, rod .....Ch. 8, p. 104, 105, 106, 107, 108
Chemical treatment ....Ch. 3, p. 35, 36
Clegg, J.D. ....Foreword, p. viii
                                Ch. 2, p.12, 14
                                Ch. 8 , p. 111
                                Ch. 11, p. 149, 150
                                Ref., p. 152
Compounds of stator elastomer ...Ch. 1, p. 7, 8, 9, 10

Conclusion .... Ch. 11, p. 149, 150, 151

Cost Comparisons of equipment ..... Ch. 2, p. 20, 21, 22, 23, 24

Coupling, coated .... Ch. 8, p. 105, 106

D

Data sheets, application .... Ch. 4, p. 54

Design explanation of PCP ...... Ch. 1, p. 3, 4, 5, 6, 7

Development needs, future .... Ch. 10, p. 140-146

Dewatering applications .... Ch. 4, p. 42

Disposal pump applications ...... Ch. 4, p. 49, 50, 51

Drilling Motors utilization of

    Moineau principle .... Ch. 1, p. 2, Ch. 10, p. 145

Drive head installation .... Ch. 5, p. 67, 68, 69, 70

Dual completion applications .... Ch. 4, p. 48

E

Efficiencies/PCP .... Ch. 2, p. 12, 13, 14

Electric motors .... Ch. 8, p. 98, 99

Environmental concerns .... Ch. 2, p. 27

F

Fluid sample .... Ch. 4, p. 52

Foreword .... p. viii

G

Generators ....Ch. 8, p. 103

H

H$_2$S Applications ....Ch. 3, p. 34, 35, Ch. 4, p. 52
Heavy oil applications ....Ch. 4, p. 43
Hein, N.W. ....Foreword, p. viii
Ch. 2, p. 12, 14
Ch. 8, p. 111
Ch. 11, p. 149, 150
Ref., p. 152
Hollow shaft drive head ....Ch. 5, p. 67
Hot Oiling Treatment ....Ch. 3, p. 36
Horizontal well applications ....Ch. 4, p. 48
Hydraulic drive units ....Ch. 8, p. 100

I

Inspection, pump ....Ch. 9, p. 127
Installation
   of stator and rotor ....Ch. 5, p. 63, 64, 65, 66, 67
   of drive head ....Ch. 5, p. 67, 68, 69, 70
Inventor/PC pump ....Ch. 1, p. 1

L

Licenses/Moineau principle patent .....Ch. 1, p. 1, 2
Lift head ....Ch. 8, p. 117, 118, 119

Low profile applications .... Ch. 4, p. 43

M

Matched or Performance Fitted Pumps Ch. 2, p. 12, 25
Ch. 4, p. 53, 56, 59
Ch. 8, p. 107
Ch. 9, p. 122, 130
Materials of construction of
  rotor and stator .... Ch. 1, p. 7
Molded-on guides .... Ch. 8, p. 105

N

Natural gas engines .... Ch. 8, p. 102

O

Offshore applications. .... Ch. 4, p. 46, 47
Oil Gravity/API limitations ... Ch. 1, p. 9, Ch. 3, p. 34

P

Patent/licenses of Moineau principle ... Ch. 1, p. 1, 2
Performance testing, pump .... Ch. 4, p. 56
Ch. 6, p. 73, 75
Ch. 9, p. 121-133
Power Savings/study of .... Ch. 2, p. 14-20
Problems and Solutions .... Ch. 7, p. 79-96
Pump off controls .... Ch. 8, p. 111-117

Pumps/Phased or tandem . . . . .Ch. 3, p. 30, 31
Pump Stages/
    applying to pressure capabilities . . . . . . .Ch. 1, p. 5

R
References . . . . .p. 152

S
Saveth, K.J. . . . . .Ch. 2, p. 15, Ref., p. 152
Separators . . . . .Ch. 8, p. 109, 110, 111
Service vehicles . . . . .Ch. 4, p. 57, 58
Slant well applications . . . . .Ch. 4, p. 47, 48
Snap-on rod guides .Ch. 8, p. 104

T
Temperature Ratings of
    stator elastomers . . . . .Ch. 1, p. 7-10
Torque Specifications, tubing . . . . .Ch. 5, p. 64

W
Water Flood applications . . . . .Ch. 4, p. 42
Water Sampling . . . . .Ch. 4, p. 49